Manufacturing Engineering and Technology

Manufacturing Engineering and Technology

Edited by

Levi O'Brein

Larsen & Keller
www.larsen-keller.com

Manufacturing Engineering and Technology
Edited by Levi O'Brein
ISBN: 978-1-63549-171-5 (Hardback)

Larsen & Keller

Published by Larsen and Keller Education,
5 Penn Plaza,
19th Floor,
New York, NY 10001, USA

Cataloging-in-Publication Data

Manufacturing engineering and technology / edited by Levi O'Brein.
p. cm.
Includes bibliographical references and index.
ISBN 978-1-63549-171-5
1. Production engineering. 2. Manufacturing processes. 3. Manufactures--Technological innovations. 4. Industrial engineering. I. O'Brein, Levi.
TS183 .M36 2017
670--dc23

Printed and bound in China.

For more information regarding Larsen and Keller Education and its products, please visit the publisher's website www.larsen-keller.com

Table of Contents

Preface

The book aims to shed light on some of the unexplored aspects of manufacturing engineering. It discusses in detail the different techniques and applications used in this field. Manufacturing engineering refers to the practice of researching, designing, constructing and developing technology like machines, systems, equipments, tools, etc. It is mainly used to turn raw materials into useful products. This book is a compilation of chapters that discuss the most vital concepts in the field of manufacturing engineering. Some of the diverse topics covered in it address the varied branches that fall under this category. Different approaches, evaluations and methodologies have been included in this text. Coherent flow of topics, student-friendly language and extensive use of examples make this textbook an invaluable source of knowledge.

A foreword of all chapters of the book is provided below:

Chapter 1 - The branch of engineering which studies manufacturing practices and methods is known as manufacturing engineering. The major concern of this subject is to convert raw materials into new products in ways which are economically efficient. This chapter will provide an integrated understanding of manufacturing engineering; **Chapter 2 -** The essential elements of manufacturing engineering are operations management, material requirements planning, process layout, production line, assembly line etc. Operation management concerns itself with designing and controlling the activities related to production. The text strategically encompasses and incorporates all the major elements of manufacturing engineering, providing a complete understanding; **Chapter 3 -** Mechanics, kinematics and technical drawing are some of the subdisciplines of manufacturing engineering. The text discusses the methods of manufacturing engineering and its subdisciplines in a critical manner providing key analysis to the subject matter; **Chapter 4 -** Power looms are looms that can be controlled mechanically whereas automation is the use of control systems for using equipment such as machinery, boilers and steering. Some other technologies explained in this section are conveyor system, conveyor belt, computer-aided manufacturing, computer-integrated manufacturing, pilot plant etc. The aim of this chapter is to explore all the technologies related to manufacturing; **Chapter 5 -** Machining is the procedure in which raw material is cut into a particular shape and then is controlled by a material removal process. The various processes of machining elucidated in the text are hobbing and milling. The other tools explained in the section are shapers, planers and grinding machines. The chapter closely examines the concept of machining and provides an extensive understanding of the subject; **Chapter 6 -** Packaging engineering is a wide topic, from design conceptualization to product placement. Some of the aspects of packaging engineering that have been dealt with in this text are active packaging, label dispenser, luxury packaging, mandatory labelling and shelf- ready packaging. This section has been carefully written to provide an easy understanding of the varied facets of packaging engineering; **Chapter 7 -** Distribution centers are centers that are stocked with products for distribution to wholesalers or consumers directly. Some of the storage mechanisms in the distribution center are automated storage and retrieval system, intermodal container, pallet, pallet racking, bulk boxes and crates. This text will introduce the reader to all the significant aspects of distribution centers.

At the end, I would like to thank all the people associated with this book devoting their precious time and providing their valuable contributions to this book. I would also like to express my gratitude to my fellow colleagues who encouraged me throughout the process.

Editor

1

Introduction to Manufacturing Engineering

The branch of engineering which studies manufacturing practices and methods is known as manufacturing engineering. The major concern of this subject is to convert raw materials into new products in ways which are economically efficient. This chapter will provide an integrated understanding of manufacturing engineering.

Manufacturing Engineering

Manufacturing engineering is a discipline of engineering dealing with various manufacturing sciences and practices including the research, design and development of systems, processes, machines, tools, and equipment. The manufacturing engineer's primary focus is to turn raw materials into a new or updated product in the most economic, efficient, and effective way possible.

Overview

This field also deals with the integration of different facilities and systems for producing quality products (with optimal expenditure) by applying the principles of physics and the results of manufacturing systems studies, such as the following:

• Craft or Guild • Putting-out system • British factory system • American system of manufacturing • Soviet collectivism in manufacturing • Mass production	• Computer integrated manufacturing • Computer-aided technologies in manufacturing • Just in time manufacturing • Lean manufacturing • Flexible manufacturing • Mass customization	• Agile manufacturing • Rapid manufacturing • Prefabrication • Ownership • Fabrication • Publication

A set of six-axis robots used for welding.

Manufacturing engineers develop and create physical artifacts, production processes, and technology. It is a very broad area which includes the design and development of products. Manufacturing engineering is considered to be a subdiscipline of industrial engineering/systems engineering and has very strong overlaps with mechanical engineering. Manufacturing engineers' success or failure directly impacts the advancement of technology and the spread of innovation. This field of manufacturing engineering emerged from tool and die discipline in the early 20th century. It expanded greatly from the 1960s when industrialized countries introduced factories with:

1. Numerical control machine tools and automated systems of production.
2. Advanced statistical methods of quality control: These factories were pioneered by the American electrical engineer William Edwards Deming, who was initially ignored by his home country. The same methods of quality control later turned Japanese factories into world leaders in cost-effectiveness and production quality.
3. Industrial robots on the factory floor, introduced in the late 1970s: These computer-controlled welding arms and grippers could perform simple tasks such as attaching a car door quickly and flawlessly 24 hours a day. This cut costs and improved production speed.

History

The history of manufacturing engineering can be traced to factories in the mid 19th century USA and 18th century UK. Although large home production sites and workshops were established in ancient China, ancient Rome and the Middle East, the Venice Arsenal provides one of the first examples of a factory in the modern sense of the word. Founded in 1104 in the Republic of Venice several hundred years before the Industrial Revolution, this factory mass-produced ships on assembly lines using manufactured parts. The Venice Arsenal apparently produced nearly one ship every day and, at its height, employed 16,000 people.

Many historians regard Matthew Boulton's Soho Manufactory (established in 1761 in Birmingham) as the first modern factory. Similar claims can be made for John Lombe's silk mill in Derby (1721), or Richard Arkwright's Cromford Mill (1771). The Cromford Mill was purpose-built to accommodate the equipment it held and to take the material through the various manufacturing processes.

Ford assembly line, 1913.

One historian, Murno Gladst, contends that the first factory was in Potosí. The Potosi factory took advantage of the abundant silver that was mined nearby and processed silver ingot slugs into coins.

British colonies in the 19th century built factories simply as buildings where a large number of workers gathered to perform hand labor, usually in textile production. This proved more efficient for the administration and distribution of materials to individual workers than earlier methods of manufacturing, such as cottage industries or the putting-out system.

Cotton mills used inventions such as the steam engine and the power loom to pioneer the industrial factories of the 19th century, where precision machine tools and replaceable parts allowed greater efficiency and less waste. This experience formed the basis for the later studies of manufacturing engineering. Between 1820 and 1850, non-mechanized factories supplanted traditional artisan shops as the predominant form of manufacturing institution.

Henry Ford further revolutionized the factory concept and thus manufacturing engineering in the early 20th century with the innovation of mass production. Highly specialized workers situated alongside a series of rolling ramps would build up a product such as (in Ford's case) an automobile. This concept dramatically decreased production costs for virtually all manufactured goods and brought about the age of consumerism.

Modern Developments

Modern manufacturing engineering studies include all intermediate processes required for the production and integration of a product's components.

Some industries, such as semiconductor and steel manufacturers use the term "fabrication" for these processes.

KUKA industrial robots being used at a bakery for food production

Automation is used in different processes of manufacturing such as machining and welding. Automated manufacturing refers to the application of automation to produce goods in a factory. The main advantages of automated manufacturing for the manufacturing process are realized with effective implementation of automation and include: higher consistency and quality, reduction of lead times, simplification of production, reduced handling, improved work flow, and improved worker morale.

Robotics is the application of mechatronics and automation to create robots, which are often used in manufacturing to perform tasks that are dangerous, unpleasant, or repetitive. These robots may be of any shape and size, but all are preprogrammed and interact physically with the world. To create a robot, an engineer typically employs kinematics (to determine the robot's range of motion) and mechanics (to determine the stresses within the robot). Robots are used extensively in manufacturing engineering.

Robots allow businesses to save money on labor, perform tasks that are either too dangerous or too precise for humans to perform economically, and to ensure better quality. Many companies employ assembly lines of robots, and some factories are so robotized that they can run by themselves. Outside the factory, robots have been employed in bomb disposal, space exploration, and many other fields. Robots are also sold for various residential applications.

Education

Certification Programs

Manufacturing engineers possess a bachelor's degree in engineering with a major in manufacturing engineering. The length of study for such a degree is usually four to five years followed by five more years of professional practice to qualify as a professional engineer. Working as a manufacturing engineering technologist involves a more applications-oriented qualification path.

Academic degrees for manufacturing engineers are usually the Bachelor of Engineering, [BE] or [BEng], and the Bachelor of Science, [BS] or [BSc]. For manufacturing technologists the required degrees are Bachelor of Technology [B.TECH] or Bachelor of Applied Science [BASc] in Manufacturing, depending upon the university. Master's degrees in engineering manufacturing include Master of Engineering [ME] or [MEng] in Manufacturing, Master of Science [M.Sc] in Manufacturing Management, Master of Science [M.Sc] in Industrial and Production Management, and Master of Science [M.Sc] as well as Master of Engineering [ME] in Design, which is a subdiscipline of manufacturing. Doctoral [PhD] or [DEng] level courses in manufacturing are also available depending on the university.

The undergraduate degree curriculum generally includes courses in physics, mathematics, computer science, project management, and specific topics in mechanical and manufacturing engineering. Initially such topics cover most, if not all, of the subdisciplines of manufacturing engineering. Students then choose to specialize in one or more subdisciplines towards the end of their degree work.

Syllabus

The foundational curriculum for a bachelor's degree in manufacturing engineering is very similar to that for mechanical engineering, and includes:

- Statics and dynamics
- Strength of materials and solid mechanics
- Instrumentation and measurement
- Applied thermodynamics, heat transfer, energy conversion, and HVAC
- Fluid mechanics and fluid dynamics
- Mechanism design (including kinematics and dynamics)
- Manufacturing technology or processes
- Hydraulics and pneumatics
- Mathematics - in particular, calculus, differential equations, statistics, and linear algebra.
- Engineering design and graphics
- Circuit Analysis

- Lean manufacturing
- Mechatronics and control theory
- Automation and reverse engineering
- Quality assurance and control
- Material science
- Drafting, CAD (including solid modeling), and CAM, etc.

A bachelor's degree in these two areas will typically differ only by a few specialized classes, although the mechanical engineering degree requires more mathematics expertise.

Manufacturing Engineering Certification

Certification and licensure:

In some countries, "professional engineer" is the term for registered or licensed engineers who are permitted to offer their professional services directly to the public. Professional Engineer, abbreviated (PE - USA) or (PEng - Canada), is the designation for licensure in North America. In order to qualify for this license, a candidate needs a bachelor's degree from an ABET recognized university in the USA, a passing score on a state examination, and four years of work experience usually gained via a structured internship. In the USA, more recent graduates have the option of dividing this licensure process into two segments. The Fundamentals of Engineering (FE) exam is often taken immediately after graduation and the Principles and Practice of Engineering exam is taken after four years of working in a chosen engineering field.

Society of Manufacturing Engineers (SME) certifications (USA):

The SME administers qualifications specifically for the manufacturing industry. These are not degree level qualifications and are not recognized at the professional engineering level. The following discussion deals with qualifications in the USA only. Qualified candidates for the Certified Manufacturing Technologist Certificate (CMfgT) must pass a three-hour, 130-question multiple-choice exam. The exam covers math, manufacturing processes, manufacturing management, automation, and related subjects. Additionally, a candidate must have at least four years of combined education and manufacturing-related work experience.

Certified Manufacturing Engineer (CMfgE) is an engineering qualification administered by the Society of Manufacturing Engineers, Dearborn, Michigan, USA. Candidates qualifying for a Certified Manufacturing Engineer credential must pass a four-hour, 180 question multiple-choice exam which covers more in-depth topics than does the CMfgT exam. CMfgE candidates must also have eight years of combined education and manufacturing-related work experience, with a minimum of four years of work experience.

Certified Engineering Manager (CEM). The Certified Engineering Manager Certificate is also designed for engineers with eight years of combined education and manufacturing experience. The test is four hours long and has 160 multiple-choice questions. The CEM certification exam covers business processes, teamwork, responsibility, and other management-related categories.

Modern Tools

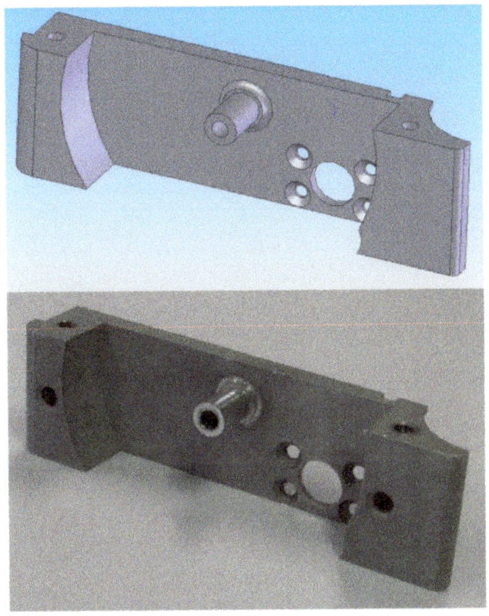

CAD model and CNC machined part

Many manufacturing companies, especially those in industrialized nations, have begun to incorporate computer-aided engineering (CAE) programs into their existing design and analysis processes, including 2D and 3D solid modeling computer-aided design (CAD). This method has many benefits, including easier and more exhaustive visualization of products, the ability to create virtual assemblies of parts, and ease of use in designing mating interfaces and tolerances.

Other CAE programs commonly used by product manufacturers include product life cycle management (PLM) tools and analysis tools used to perform complex simulations. Analysis tools may be used to predict product response to expected loads, including fatigue life and manufacturability. These tools include finite element analysis (FEA), computational fluid dynamics (CFD), and computer-aided manufacturing (CAM).

Using CAE programs, a mechanical design team can quickly and cheaply iterate the design process to develop a product that better meets cost, performance, and other constraints. No physical prototype need be created until the design nears completion, allowing hundreds or thousands of designs to be evaluated, instead of relatively few. In addition, CAE analysis programs can model complicated physical phenomena which cannot be solved by hand, such as viscoelasticity, complex contact between mating parts, or non-Newtonian flows.

Just as manufacturing engineering is linked with other disciplines, such as mechatronics, multidisciplinary design optimization (MDO) is also being used with other CAE programs to automate and improve the iterative design process. MDO tools wrap around existing CAE processes, allowing product evaluation to continue even after the analyst goes home for the day. They also utilize sophisticated optimization algorithms to more intelligently explore possible designs, often finding better, innovative solutions to difficult multidisciplinary design problems.

Subdisciplines

Mechanics

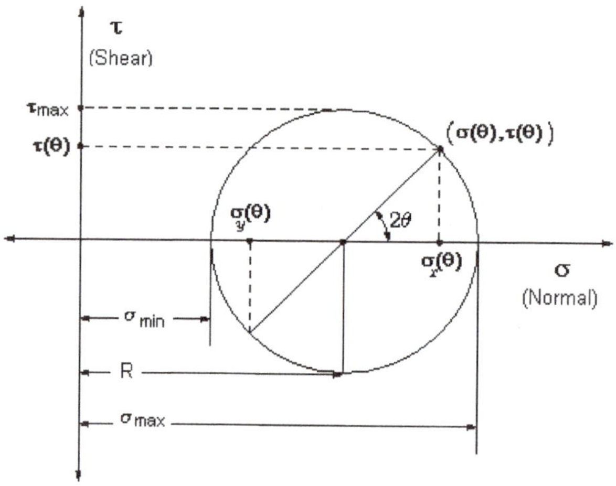

Mohr's circle, a common tool to study stresses in a mechanical element

Mechanics, in the most general sense, is the study of forces and their effects on matter. Typically, engineering mechanics is used to analyze and predict the acceleration and deformation (both elastic and plastic) of objects under known forces (also called loads) or stresses. Subdisciplines of mechanics include:

- Statics, the study of non-moving bodies under known loads
- Dynamics (or kinetics), the study of how forces affect moving bodies
- Mechanics of materials, the study of how different materials deform under various types of stress
- Fluid mechanics, the study of how fluids react to forces
- Continuum mechanics, a method of applying mechanics that assumes that objects are continuous (rather than discrete)

If the engineering project were to design a vehicle, statics might be employed to design the frame of the vehicle in order to evaluate where the stresses will be most intense. Dynamics might be used when designing the car's engine to evaluate the forces in the pistons and cams as the engine cycles. Mechanics of materials might be used to choose appropriate materials for the manufacture of the frame and engine. Fluid mechanics might be used to design a ventilation system for the vehicle or to design the intake system for the engine.

Kinematics

Kinematics is the study of the motion of bodies (objects) and systems (groups of objects), while ignoring the forces that cause the motion. The movement of a crane and the oscillations of a piston in an engine are both simple kinematic systems. The crane is a type of open kinematic chain, while the piston is part of a closed four-bar linkage. Engineers typically use kinematics in the design and analysis of mechanisms. Kinematics can be used to find the possible range of motion for a given mechanism, or, working in reverse, can be used to design a mechanism that has a desired range of motion.

Drafting

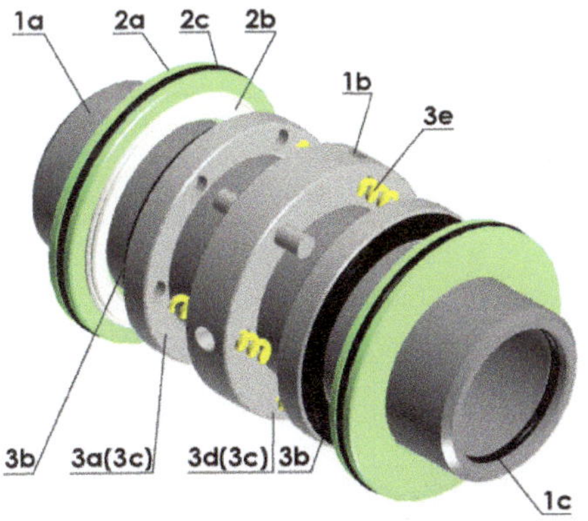

A CAD model of a mechanical double seal

Drafting or technical drawing is the means by which manufacturers create instructions for manufacturing parts. A technical drawing can be a computer model or hand-drawn schematic showing all the dimensions necessary to manufacture a part, as well as assembly notes, a list of required materials, and other pertinent information. A U.S engineer or skilled worker who creates technical drawings may be referred to as a drafter or draftsman. Drafting has historically been a two-dimensional process, but computer-aided design (CAD) programs now allow the designer to create in three dimensions.

Instructions for manufacturing a part must be fed to the necessary machinery, either manually, through programmed instructions, or through the use of a computer-aided manufacturing (CAM) or combined CAD/CAM program. Optionally, an engineer may also manually manufacture a part using the technical drawings, but this is becoming an increasing rarity with the advent of computer numerically controlled (CNC) manufacturing. Engineers primarily manufacture parts manually in the areas of applied spray coatings, finishes, and other processes that cannot economically or practically be done by a machine.

Drafting is used in nearly every subdiscipline of mechanical and manufacturing engineering, and by many other branches of engineering and architecture. Three-dimensional models created using CAD software are also commonly used in finite element analysis (FEA) and computational fluid dynamics (CFD).

Machine Tools and Metal Fabrication

Machine tools employ some sort of tool that does the cutting or shaping. All machine tools have some means of constraining the workpiece and provide a guided movement of the parts of the machine. Metal fabrication is the building of metal structures by cutting, bending, and assembling processes.

Computer Integrated Manufacturing

Computer-integrated manufacturing (CIM) is the manufacturing approach of using computers to control the entire production process. Computer-integrated manufacturing is used in automotive, aviation, space, and ship building industries.

Mechatronics

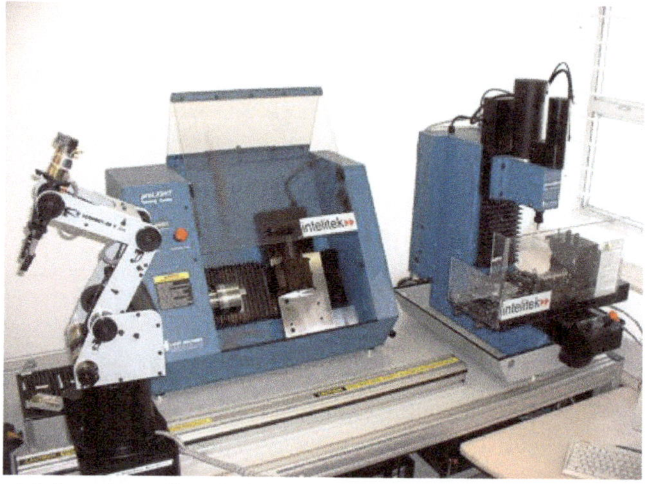

Training FMS with learning robot SCORBOT-ER 4u, workbench CNC mill and CNC lathe

Mechatronics is an engineering discipline that deals with the convergence of electrical, mechanical and manufacturing systems. Such combined systems are known as electromechanical systems and are widespread. Examples include automated manufacturing systems, heating, ventilation and air-conditioning systems, and various aircraft and automobile subsystems.

The term mechatronics is typically used to refer to macroscopic systems, but futurists have predicted the emergence of very small electromechanical devices. Already such small devices, known as Microelectromechanical systems (MEMS), are used in automobiles to initiate the deployment of airbags, in digital projectors to create sharper images, and in inkjet printers to create nozzles for high-definition printing. In the future it is hoped that such devices will be used in tiny implantable medical devices and to improve optical communication.

Textile Engineering

Textile engineering courses deal with the application of scientific and engineering principles to the design and control of all aspects of fiber, textile, and apparel processes, products, and machinery. These include natural and man-made materials, interaction of materials with machines, safety and health, energy conservation, and waste and pollution control. Additionally, students are given experience in plant design and layout, machine and wet process design and improvement, and designing and creating textile products. Throughout the textile engineering curriculum, students take classes from other engineering and disciplines including: mechanical, chemical, materials and industrial engineering.

Advanced Composite Materials

Advanced composite materials (engineering) (ACMs) are also known as advanced polymer matrix composites. These are generally characterized or determined by unusually high strength fibres with unusually high stiffness, or modulus of elasticity characteristics, compared to other materials, while bound together by weaker matrices. Advanced composite materials have broad, proven applications, in the aircraft, aerospace, and sports equipment sectors. Even more specifically ACMs are very attractive for aircraft and aerospace structural parts. Manufacturing ACMs is a multibillion-dollar industry worldwide. Composite products range from skateboards to components of the space shuttle. The industry can be generally divided into two basic segments, industrial composites and advanced composites.

Employment

Manufacturing engineering is just one facet of the engineering industry. Manufacturing engineers enjoy improving the production process from start to finish. They have the ability to keep the whole production process in mind as they focus on a particular portion of the process. Successful students in manufacturing engineering degree programs are inspired by the notion of starting with a natural resource, such as a block of wood, and ending with a usable, valuable product, such as a desk, produced efficiently and economically.

Manufacturing engineers are closely connected with engineering and industrial design efforts. Examples of major companies that employ manufacturing engineers in the United States include General Motors Corporation, Ford Motor Company, Chrysler, Boeing, Gates Corporation and Pfizer. Examples in Europe include Airbus, Daimler, BMW, Fiat, Navistar International, and Michelin Tyre.

Industries where manufacturing engineers are generally employed include:

- Aerospace industry
- Automotive industry
- Chemical industry
- Computer industry
- Food processing industry
- Garment industry
- Pharmaceutical industry
- Pulp and paper industry
- Toy industry

Frontiers of Research

Flexible Manufacturing Systems

A typical FMS system

A flexible manufacturing system (FMS) is a manufacturing system in which there is some amount of flexibility that allows the system to react to changes, whether predicted or unpredicted. This flexibility is generally considered to fall into two categories, both of which have numerous subcategories. The first category, machine flexibility, covers the system's ability to be changed to produce new product types and the ability to change the order of operations executed on a part. The second category, called routing flexibility, consists of the ability to use multiple machines to perform the same operation on a part, as well as the system's ability to absorb large-scale changes, such as in volume, capacity, or capability.

Most FMS systems comprise three main systems. The work machines, which are often automated CNC machines, are connected by a material handling system to optimize parts flow, and to a central control computer, which controls material movements and machine flow. The main advantages of an FMS is its high flexibility in managing manufacturing resources like time and effort in order to manufacture a new product. The best application of an FMS is found in the production of small sets of products from a mass production.

Computer Integrated Manufacturing

Computer-integrated manufacturing (CIM) in engineering is a method of manufacturing in which the entire production process is controlled by computer. Traditionally separated process methods are joined through a computer by CIM. This integration allows the processes to exchange information and to initiate actions. Through this integration, manufacturing can be faster and less error-prone, although the main advantage is the ability to create automated manufacturing processes. Typically CIM relies on closed-loop control processes based on real-time input from sensors. It is also known as flexible design and manufacturing.

Friction Stir Welding

Close-up view of a friction stir weld tack tool

Friction stir welding was discovered in 1991 by The Welding Institute (TWI). This innovative steady state (non-fusion) welding technique joins previously un-weldable materials, including several aluminum alloys. It may play an important role in the future construction of airplanes, potentially replacing rivets. Current uses of this technology to date include: welding the seams of the aluminum main space shuttle external tank, the Orion Crew Vehicle test article, Boeing Delta II and Delta IV Expendable Launch Vehicles and the SpaceX Falcon 1 rocket; armor plating for amphibious assault ships; and welding the wings and fuselage panels of the new Eclipse 500 aircraft from Eclipse Aviation, among an increasingly growing range of uses.

Other areas of research are Product Design, MEMS (Micro-Electro-Mechanical Systems), Lean Manufacturing, Intelligent Manufacturing Systems, Green Manufacturing, Precision Engineering, Smart Materials, etc.

Manufacturing

Manufacturing is the value added production of merchandise for use or sale using labour and machines, tools, chemical and biological processing, or formulation. The term may refer to a range of human activity, from handicraft to high tech, but is most commonly applied to industrial production, in which raw materials are transformed into finished goods on a large scale. Such finished goods may be sold to other manufacturers for the production of other, more complex products, such as aircraft, household appliances or automobiles, or sold to wholesalers, who in turn sell them to retailers, who then sell them to end users and consumers.

Manufacturing or manufacturing process are the steps through which raw materials are transformed into a final product. The manufacturing process begins with the creation of the materials from which the design is made. These materials are then modified through manufacturing processes to become the required part.

Manufacturing takes turns under all types of economic systems. In a free market economy, manufacturing is usually directed toward the mass production of products for sale to consumers at a profit. In a collectivist economy, manufacturing is more frequently directed by the state to supply a centrally planned economy. In mixed market economies, manufacturing occurs under some degree of government regulation.

Modern manufacturing includes all intermediate processes required for the production and integration of a product's components. Some industries, such as semiconductor and steel manufacturers use the term fabrication instead.

The manufacturing sector is closely connected with engineering and industrial design. Examples of major manufacturers in North America include General Motors Corporation, General Electric, Procter & Gamble, General Dynamics, Boeing, Pfizer, and Precision Castparts. Examples in Europe include Volkswagen Group, Siemens, and Michelin. Examples in Asia include Sony, Huawei, Lenovo, Toyota, Samsung, and Bridgestone.

History and Development

Assembly of Section 41 of a Boeing 787 Dreamliner

A female industrial worker amidst heavy steel semi-products (KINEX BEARINGS, Bytča, Slovakia, c. 1995–2000)

A modern automobile assembly line.

- In its earliest form, manufacturing was usually carried out by a single skilled artisan with assistants. Training was by apprenticeship. In much of the pre-industrial world, the guild system protected the privileges and trade secrets of urban artisans.

- Before the Industrial Revolution, most manufacturing occurred in rural areas, where household-based manufacturing served as a supplemental subsistence strategy to agriculture (and continues to do so in places). Entrepreneurs organized a number of manufacturing households into a single enterprise through the putting-out system.
- Toll manufacturing is an arrangement whereby a first firm with specialized equipment processes raw materials or semi-finished goods for a second firm.

Manufacturing Systems: Changes in Methods of Manufacturing

- Agile manufacturing
- American system of manufacturing
- British factory system of manufacturing
- Craft or guild system
- Fabrication
- Flexible manufacturing
- Just-in-time manufacturing
- Lean manufacturing
- Mass customization (2000s) - 3D printing, design-your-own web sites for sneakers, fast fashion
- Mass production
- Ownership
- Packaging and labeling
- Prefabrication
- Putting-out system
- Rapid manufacturing
- Reconfigurable manufacturing system
- Soviet collectivism in manufacturing

Industrial Policy

Economics of Manufacturing

According to some economists, manufacturing is a wealth-producing sector of a country, whereas a service sector tends to be wealth-consuming. Emerging technologies have provided some new growth in advanced manufacturing employment opportunities in the Manufacturing Belt in the United States. Manufacturing provides important material support for national infrastructure and for national defense.

On the other hand, mwost manufacturing may involve significant social and environmental costs. The clean-up costs of hazardous waste, for example, may outweigh the benefits of a product that creates it. Hazardous materials may expose workers to health risks. These costs are now well known and there is effort to address them by improving efficiency, reducing waste, using industrial symbiosis, and eliminating harmful chemicals. The increased use of technologies such as 3D printing also offer the potential to reduce the environmental impact of producing finished goods through distributed manufacturing.

The negative costs of manufacturing can also be addressed legally. Developed countries regulate manufacturing activity with labor laws and environmental laws. Across the globe, manufacturers can be subject to regulations and pollution taxes to offset the environmental costs of manufacturing activities. Labor unions and craft guilds have played a historic role in the negotiation of worker rights and wages. Environment laws and labor protections that are available in developed nations may not be available in the third world. Tort law and product liability impose additional costs on manufacturing. These are significant dynamics in the ongoing process, occurring over the last few decades, of manufacture-based industries relocating operations to "developing-world" economies where the costs of production are significantly lower than in "developed-world" economies.

Manufacturing and Investment

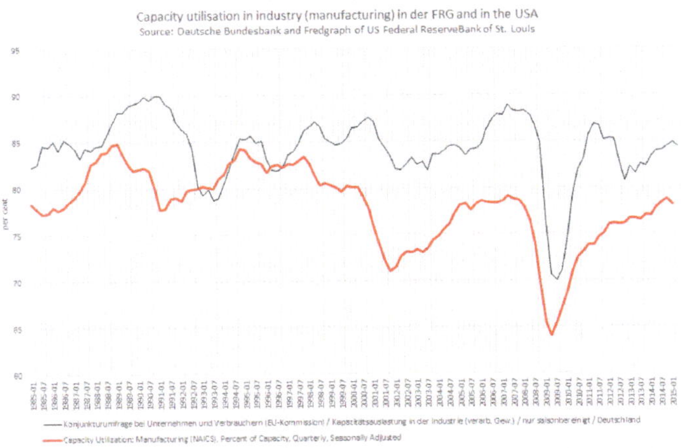

Capacity utilization in manufacturing in the FRG and in the USA

Surveys and analyses of trends and issues in manufacturing and investment around the world focus on such things as:

- The nature and sources of the considerable variations that occur cross-nationally in levels of manufacturing and wider industrial-economic growth;

- Competitiveness; and
- Attractiveness to foreign direct investors.

In addition to general overviews, researchers have examined the features and factors affecting particular key aspects of manufacturing development. They have compared production and investment in a range of Western and non-Western countries and presented case studies of growth and performance in important individual industries and market-economic sectors.

On June 26, 2009, Jeff Immelt, the CEO of General Electric, called for the United States to increase its manufacturing base employment to 20% of the workforce, commenting that the U.S. has outsourced too much in some areas and can no longer rely on the financial sector and consumer spending to drive demand. Further, while U.S. manufacturing performs well compared to the rest of the U.S. economy, research shows that it performs poorly compared to manufacturing in other high-wage countries. A total of 3.2 million – one in six U.S. manufacturing jobs – have disappeared between 2000 and 2007. In the UK, EEF the manufacturers organisation has led calls for the UK economy to be rebalanced to rely less on financial services and has actively promoted the manufacturing agenda.

Countries by Manufacturing Output Using the Most Recent Known Data

Data is provided by Worldbank. It shows the total value of manufacturing in US dollars for its noted year.

Rank	Country/Region	Millions of $US	Year
	World	**12,578,627**	**2014**
1	China	3,713,300	2014
	European Union	2,566,070	2014
2	United States	2,068,080	2014
	Eurozone	1,946,857	2014
3	Japan	850,902	2014
4	Germany	787,503	2014
5	South Korea	389,582	2014
6	India	321,721	2014
7	Italy	296,611	2014
8	France	283,664	2014

Rank	Country/Region	Millions of $US	Year
9	United Kingdom	282,675	2014
10	Russia	248,481	2014
11	Brazil	218,799	2014
12	Mexico	216,773	2014
13	Indonesia	186,744	2014
14	Spain	166,594	2014
15	Canada	162,074	2014
16	Switzerland	128,881	2014
17	Turkey	126,365	2014
18	Thailand	112,214	2014
19	Netherlands	95,683	2014
20	Australia	93,461	2016

Manufacturing Processes

- List of manufacturing processes
- Manufacturing Process Management

Theories

- Taylorism/Scientific management
- Fordism

Control

- Management
    - List of management topics
    - Total Quality Management
- Quality control
    - Six Sigma

Process Manufacturing

Process manufacturing is the branch of manufacturing that is associated with formulas and manufacturing recipes, and can be contrasted with discrete manufacturing, which is concerned with discrete units, bills of materials and the assembly of components.

Process manufacturing is common in the food, beverage, chemical, pharmaceutical, consumer packaged goods, and biotechnology industries. In process manufacturing, the relevant factors are ingredients, not parts; formulas, not bills of materials; and bulk materials rather than individual units. Although there is invariably cross-over between the two branches of manufacturing, the major contents of the finished product and the majority of the resource intensity of the production process generally allow manufacturing systems to be classified as one or the other. For example, a bottle of juice is a discrete item, but juice is process manufactured. The plastic used in injection moulding is process manufactured, but the components it is shaped into are generally discrete, and subject to further assembly.

Formulation

Formulation is a simple concept, but it is often incorrectly equated with a bill of materials. Formulation specifies the ingredients and the amounts (e.g., pounds, gallons, liters) needed to make the product. The first thing to recognize is that to be able to work with a formula, the units of measure must correspond; a flexible unit of measure conversion engine running under an ERP software cover is needed. Furthermore, conversion rules must be specified to account for the unique requirements of the business in question.

The proportions of ingredients in a formula also highlight the need for another feature, namely scalability. A formula to make 500 liters of a chemical must be scalable to make 250 liters or 1,000 liters. Another aspect of scalability is that it makes possible manufacturing based on how much of an ingredient is available. An example will illustrate this point. If you are making a car and only have two of the required four tires, you cannot make half a car. In other words, you must have all the parts in the required quantities to make the finished product; they are not scalable. But in process manufacturing, if you want to make 1,000 gallons of soda and you only have 500 gallons of the required 1,000 gallons of carbonated water, you have the option of making half as much soda. In process manufacturing you can make as much of a finished product as is specified in the formula for the smallest quantity in stock of one of the ingredients.

Packaging

A packaging recipe is similar to a formula, but rather than describing the proportion of ingredients, it specifies how the finished product gets to its final assembly. A packaging recipe addresses such things as containers, labels, corrugated cartons, and shrink-wrapping. In process manufacturing, the finished product is usually produced in bulk, but is rarely delivered in bulk form to the customer. For example, the beverage manufacturer makes soda in batches of thousands of gallons. However, a consumer purchases soda in 12-ounce aluminum cans, or in 16-ounce plastic bottles, or in 1-liter bottles. And a restaurateur may have the option of getting a 5- or 50-gallon metal container with the beverage in syrup form, so that carbonated water can be added later.

Why is this concept important? Compare how often Coca Cola changes the formula for Coke with how often the packaging is changed. If the formula and packaging recipes are linked, then every time the packaging changes, the formula would need modification. Likewise, when the formula is changed, all of the packaging recipes would have to be changed. This increases maintenance costs and chances for error. In process manufacturing, the formula for making the product and the recipe for packaging the product exist in separate structures to reduce the ongoing maintenance function.

In the production cycle, a work order is issued to make the product in bulk. Separate pack orders are issued to signify how the bulk material is to be containerized and shipped to the customer. This is important in process industries which make "brite" stock or private labels. For example, large grocery chains sell products, such as soups, soda, and meats, under their own brand names, hence "private labels". But these chains do not have their own manufacturing plants; they contract for these products. In the case of soups, process manufacturers create and warehouse nondescript, unlabeled (hence "brite") aluminum cans of soup. (Since the cans are filled, sealed, and then cooked under pressure, their shelf life is long.)

By separating the product formula from a packaging recipe, a production order can be issued to make and store the cans of soup and later, when the customer is ready to order soup, a work order can be issued to label the cans according to customer specifications before they are shipped to the store. Thus segregation of the formula and pack recipe makes the world of process manufacturing efficient and effective.

Process Manufacturing Software

Just like the products that they produce, discrete and processing manufacturing software have different focal points and solve different problems. For the same reason that the proverbial square peg does not fit in the round hole, software geared toward discrete, or even hybrid manufacturing will not work smoothly in a process manufacturing setting. Even process manufacturing software alone needs to be tailored to a particular business context. Critical aspects such as formulation, routing, ingredients, unit of measure, lot trace-ability and product & implementation pricing must be evaluated relative to the business.

References

- Manufacturing & Investment Around The World: An International Survey Of Factors Affecting Growth & Performance, ISR Publications/Google Books, revised second edition, 2002. ISBN 978-0-906321-25-6.
- Research, Industrial Systems (2002-05-20). "Manufacturing and Investment Around the World: An International Survey of Factors Affecting Growth and Performance". ISBN 978-0-906321-25-6.

2

Essential Elements of Manufacturing Engineering

The essential elements of manufacturing engineering are operations management, material requirements planning, process layout, production line, assembly line etc. Operation management concerns itself with designing and controlling the activities related to production. The text strategically encompasses and incorporates all the major elements of manufacturing engineering, providing a complete understanding.

Operations Management

Operations management is an area of management concerned with designing and controlling the process of production and redesigning business operations in the production of goods or services. It involves the responsibility of ensuring that business operations are efficient in terms of using as few resources as needed and effective in terms of meeting customer requirements. It is concerned with managing the process that converts inputs (in the forms of raw materials, labor, and energy) into outputs (in the form of goods and/or services). The relationship of operations management to senior management in commercial contexts can be compared to the relationship of line officers to highest-level senior officers in military science. The highest-level officers shape the strategy and revise it over time, while the line officers make tactical decisions in support of carrying out the strategy. In business as in military affairs, the boundaries between levels are not always distinct; tactical information dynamically informs strategy, and individual people often move between roles over time.

Ford Motor car assembly line: the classical example of a manufacturing production system.

Post office queue. Operations management studies both manufacturing and services.

According to the United States Department of Education, operations management is the field concerned with managing and directing the physical and/or technical functions of a firm or organization, particularly those relating to development, production, and manufacturing. Operations management programs typically include instruction in principles of general management, manufacturing and production systems, factory management, equipment maintenance management, production control, industrial labor relations and skilled trades supervision, strategic manufacturing policy, systems analysis, productivity analysis and cost control, and materials planning.

History

The history of production and operation systems began around 5000 B.C. when Sumerian priests developed the ancient system of recording inventories, loans, taxes, and business transactions. The next major historical application of operation systems occurred in 4000 B.C. It was during this time that the Egyptians started using planning, organization, and control in large projects such as the construction of the pyramids. By 1100 B.C., labor was being specialized in China; by about 370 B.C., Xenophon described the advantages of dividing the various operations necessary for the production of shoes among different individuals in ancient Greece.

Shoemakers, 1568

In the Middle Ages, kings and queens ruled over large areas of land. Loyal noblemen maintained large sections of the monarch's territory. This hierarchical organization in which people were divided into classes based on social position and wealth became known as the feudal system. In the feudal system, vassals and serfs produced for themselves and people of higher classes by using the ruler's land and resources. Although a large part of labor was employed in agriculture, artisans contributed to economic output and formed guilds. The guild system, operating mainly between 1100 and 1500, consisted of two types: merchant guilds, who bought and sold goods, and craft guilds, which made goods. Although guilds were regulated as to the quality of work performed, the resulting system was rather rigid, shoemakers, for example, were prohibited from tannin hides.

The industrial revolution was facilitated by two elements: interchangeability of parts and division of labor. Division of labor has always been a feature from the beginning of civilization, the extent to which the division is carried out varied considerably depending on period and location. Compared to the Middle Ages, the Renaissance and the Age of Discovery were characterized by a greater specialization in labor, one of the characteristics of growing European cities and trade. It was in the late eighteenth century that Eli Whitney popularized the concept of interchangeability of parts when he manufactured 10,000 muskets. Up to this point in the history of manufacturing, each product (e.g. each gun) was considered a special order, meaning that parts of a given gun were fitted only for that particular gun and could not be used in other guns. Interchangeability of parts allowed the mass production of parts independent of the final products in which they would be used.

In 1883, Frederick Winslow Taylor introduced the stopwatch method for accurately measuring the time to perform each single task of a complicated job. He developed the scientific study of productivity and identifying how to coordinate different tasks to eliminate wasting of time and increase the quality of work. The next generation of scientific study occurred with the development of work sampling and predetermined motion time systems (PMTS). Work sampling is used to measure the random variable associated with the time of each task. PMTS allows the use of standard predetermined tables of the smallest body movements (e.g. turning the left wrist by 90°), and integrating them to predict the time needed to perform a simple task. PMTS has gained substantial importance due to the fact that it can predict work measurements without observing the actual work. The foundation of PMTS was laid out by the research and development of Frank B. and Lillian M. Gilbreth around 1912. The Gilbreths took advantage of taking motion pictures at known time intervals while operators were performing the given task.

The idea of the production line has been used multiple times in history prior to Henry Ford: the Venetian Arsenal (1104); Smith's pin manufacturing, in the Wealth of Nations (1776) or Brunel's Portsmouth Block Mills (1802). Ransom Olds was the first to manufacture cars using the assembly line system, but Henry Ford developed the first auto assembly system where a car chassis was moved through the assembly line by a conveyor belt while workers added components to it until the car was completed. During World War II, the growth of computing power led to further development of efficient manufacturing methods and the use of advanced mathematical and statistical tools. This was supported by the development of academic programs in industrial and systems engineering disciplines, as well as fields of operations research and management science (as multi-disciplinary fields of problem solving). While systems engineering concentrated on the broad characteristics of the relationships between inputs and outputs of generic systems, operations researchers concentrated on solving specific and focused problems. The synergy of operations research and systems engineering allowed for the realization of solving large scale and complex problems in the modern era. Recently, the development of faster and smaller computers, intelligent systems, and the World Wide Web has opened new opportunities for operations, manufacturing, production, and service systems.

Malakooti (2013) states that production and operation systems can be divided into five phases:

1. Empiricism (learning from experience)
2. Analysis (scientific management)
3. Synthesis (development of mathematical problem solving tools)
4. Isolated Systems with Single Objective (use of Integrated and Intelligent Systems, and WWW)
5. Integrated Complex Systems with Multiple Objectives (development of ecologically sound systems, environmentally sustainable systems, considering individual preferences)

Industrial Revolution

Marshall's flax mill in Holbeck. The textile industry is the prototypical example of the English industrial revolution.

Before the First industrial revolution work was mainly done through two systems: domestic system and craft guilds. In the domestic system merchants took materials to homes where artisans performed the necessary work, craft guilds on the other hand were associations of artisans which passed work from one shop to another, for example: leather was tanned by a tanner, passed to curriers, and finally arrived at shoemakers and saddlers.

The beginning of the industrial revolution is usually associated with 18th century English textile industry, with the invention of flying shuttle by John Kay in 1733, the spinning jenny by James Hargreaves in 1765, the water frame by Richard Arkwright in 1769 and the steam engine by James Watt in 1765. In 1851 at the Crystal Palace Exhibition the term American system of manufacturing was used to describe the new approach that was evolving in the United States of America which was based on two central features: interchangeable parts and extensive use of mechanization to produce them.

Henry Ford was 39 years old when he founded the Ford Motor Company in 1903, with $28,000 capital from twelve investors. The model T car was introduced in 1908, however it was not until Ford implemented the assembly line concept, that his vision of making a popular car affordable by every middle-class American citizen would be realized. The first factory in which Henry Ford used the concept of the assembly line was Highland Park (1913), he characterized the system as follows:

"The thing is to keep everything in motion and take the work to the man and not the man to the work. That is the real principle of our production, and conveyors are only one of many means to an end"

This became one the central ideas that led to mass production, one of the main elements of the Second Industrial Revolution, along with emergence of the electrical industry and petroleum industry.

Operations Management

Although productivity benefited considerably from technological inventions and division of labour, the problem of systematic measurement of performances and the calculation of these by the use of remained somewhat unexplored until Frederick Taylor, whose early work focused on developing what he called a "differential piece-rate system" and a series of experiments, measurements and formulas dealing with cutting metals and manual labor. The differential piece-rate system consisted in offering two different pay rates for doing a job: a higher rate for workers with high productivity (efficiency) and who produced high quality goods (effectiveness) and a lower rate for those who fail to achieve the standard. One of the problems Taylor believed could be solved with this system, was the problem of soldiering: faster workers reducing their production rate to that of the slowest worker. In 1911 Taylor published his "The Principles of Scientific Management", in which he characterized scientific management (also known as Taylorism) as:

1. The development of a true science;
2. The scientific selection of the worker;

3. The scientific education and development of the worker;

4. Intimate friendly cooperation between the management and the workers.

Taylor is also credited for developing stopwatch time study, this combined with Frank and Lillian Gilbreth motion study gave way to time and motion study which is centered on the concepts of standard method and standard time. Frank Gilbreth is also responsible for introducing the flow process chart in 1921. Other contemporaries of Taylor worth remembering are Morris Cooke (rural electrification in the 1920s and implementer of Taylor's principles of scientific management in the Philadelphia's Department of Public Works), Carl Barth (speed-and-feed-calculating slide rules) and Henry Gantt (Gantt chart). Also in 1910 Hugo Diemer published the first industrial engineering book: Factory Organization and Administration.

In 1913 Ford Whitman Harris published his "How Many parts to make at once" in which he presented the idea of the economic order quantity model. He described the problem as follows:

"Interest on capital tied up in wages, material and overhead sets a maximum limit to the quantity of parts which can be profitably manufactured at one time; "setup costs" on the job fix the minimum. Experience has shown one manager a way to determine the economical size of lots"

This paper inspired a large body of mathematical literature focusing on the problem of production planning and inventory control.

In 1924 Walter Shewhart introduced the control chart through a technical memorandum while working at Bell Labs, central to his method was the distinction between common cause and special cause of variation. In 1931 Shewhart published his Economic Control of Quality of Manufactured Product, the first systematic treatment of the subject of Statistical Process Control (SPC).

In the 1940s methods-time measurement (MTM) was developed by H.B. Maynard, JL Schwab and GJ Stegemerten. MTM was the first of a series of predetermined motion time systems, predetermined in the sense that estimates of time are not determined in loco but are derived from an industry standard. This was explained by its originators in a book they published in 1948 called "Method-Time Measurement".

Up to this point in history, optimization techniques were known for a very long time, from the simple methods employed by F.W.Harris to the more elaborate techniques of the calculus of variations developed by Euler in 1733 or the multipliers employed by Lagrange in 1811, and computers were slowly being developed, first as analog computers by Sir William Thomson (1872) and James Thomson (1876) moving to the eletromechanical computers of Konrad Zuse (1939 and 1941). During World War II however, the development of mathematical optimization went through a major boost with the development of the Colossus computer, the first electronic digital computer that was all programmable, and the possibility to computationally solve large linear programming problems, first by Kantorovich in 1939 working for the Soviet government and latter on in 1947 with the simplex method of Dantzig. These methods are known today as belonging to the field of operations research.

From this point on a curious development took place: while in the United States the possibility of applying the computer to business operations led to the development of management software architecture such as MRP and successive modifications, and ever more sophisticated optimization techniques and manufacturing simulation software, in post-war Japan a series of events at Toyota Motor led to the development of the Toyota Production System (TPS) and Lean Manufacturing.

In 1943, in Japan, Taiichi Ohno arrived at Toyota Motor company. Toyota evolved a unique manufacturing system centered on two complementary notions: just in time (produce only what is needed) and autonomation (automation with a human touch). Regarding JIT, Ohno was inspired by American supermarkets: workstations functioned like a supermarket shelf where the customer can get products they need, at the time they need and in the amount needed, the workstation (shelf) is then restocked. Autonomation was developed by Toyoda Sakichi in Toyoda Spinning and Weaving: an automatically activated loom that was also foolproof, that is automatically detected problems. In 1983 J.N Edwards published his “MRP and Kanban-American style” in which he described JIT goals in terms of seven zeros: zero defects, zero (excess) lot size, zero setups, zero breakdowns, zero handling, zero lead time and zero surging. This period also marks the spread of Total Quality Management (TQM) in Japan, ideas initially developed by American authors such as Deming, Juran and Armand V. Feigenbaum. TQM is a strategy for implementing and managing quality improvement on an organizational basis, this includes: participation, work culture, customer focus, supplier quality improvement and integration of the quality system with business goals. Schnonberger identified seven fundamentals principles essential to the Japanese approach:

1. Process control: SPC and worker responsibility over quality
2. Easy able -to-see quality: boards, gauges, meters, etc. and poka-yoke
3. Insistence on compliance: “quality first”
4. Line stop: stop the line to correct quality problems
5. Correcting one’s own errors: worker fixed a defective part if he produced it
6. The 100% check: automated inspection techniques and foolproof machines
7. Continual improvement: ideally zero defects

In 1987 the International Organization for Standardization (ISO), recognizing the growing importance of quality, issued the ISO 9000, a family of standards related to quality management systems. There has been some controversy though regarding the proper procedures to follow and the amount of paperwork involved.

Meanwhile, in the sixties, a different approach was developed by George W. Plossl and Oliver W. Wight, this approach was continued by Joseph Orlicky as a response to the TOYOTA Manufacturing Program which led to Material Requirements Planning (MRP) at IBM, latter gaining momentum in 1972 when the American Production and Inventory Control Society launched the "MRP Crusade". One of the key insights of this management system was the distinction between dependent demand and independent demand. Independent demand is demand which originates outside of the production system, therefore not directly controllable, and dependent demand is demand for components of final products, therefore subject to being directly controllable by management through the bill of materials, via product design. Orlicky wrote "Materials Requirement Planning" in 1975, the first hard cover book on the subject. MRP II was developed by Gene Thomas at IBM, and expanded the original MRP software to include additional production functions. Enterprise resource planning (ERP) is the modern software architecture, which addresses, besides production operations, distribution, accounting, human resources and procurement.

Recent trends in the field revolve around concepts such as:

- Business Process Re-engineering (launched by Michael Hammer in 1993): a business management strategy focusing on the analysis and design of workflows and business processes within an organization. BPR seeks to help companies radically restructure their organizations by focusing on the ground-up design of their business processes.
- Lean Manufacturing: a systemic method for the elimination of waste ("Muda") within a manufacturing process. Lean also takes into account waste created through overburden ("Muri") and waste created through unevenness in work loads ("Mura"). The term lean manufacturing was coined in the book The Machine that Changed the World.
- Six Sigma (an approach to quality developed at Motorola between 1985-1987): Six Sigma refers to control limits placed at six (6) standard deviations from the mean of a normal distribution, this became very famous after Jack Welch of General Electric launched a company-wide initiative in 1995 to adopt this set of methods. More recently, Six Sigma has included DMAIC (for improving processes) and DFSS (for designing new products and new processes)
- Reconfigurable Manufacturing Systems: a production system designed at the outset for rapid change in its structure, as well as its hardware and software components, in order to quickly adjust its production capacity and functionality within a part family in response to sudden market changes or intrinsic system change.

Topics

Production Systems

A production system comprises both the technological elements (machines and tools) and organizational behavior (division of labor and information flow). An individual production system is usually analyzed in the literature referring to a single business, therefore it's usually improper to include in a given production system the operations necessary to process goods that are obtained by purchasing or the operations carried by the customer on the sold products, the reason being simply that since businesses need to design their own production systems this then becomes the focus of analysis, modeling and decision making (also called "configuring" a production system) .

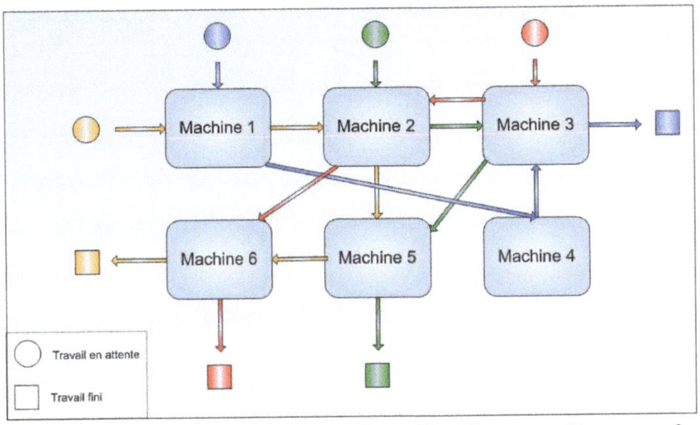

In a job shop machines are grouped by technological similarities regarding transformation processes, therefore a single shop can work very different products (in this picture four colors). Also notice that in this drawing each shop contains a single machine.

Flexible Manufacturing System: in the middle there are two rails for the shuttle to move pallets between machining centers (there are also FMS which use AGVs), in front of each machining center there is a buffer and in left we have a shelf for storing pallets. Usually in the back there is a similar system for managing the set of tools required for different machining operations.

A first possible distinction in production systems (technological classification) is between continuous process production and discrete part production (manufacturing).

- Process production means that the product undergoes physical-chemical transformations and lacks assembly operations, therefore the original raw materials can't easily be obtained from the final product, examples include: paper, cement, nylon and petroleum products.
- Part production (ex:cars and ovens) comprises both fabrication systems and assembly systems. In the first category we find job shops, manufacturing cells, flexible manufacturing systems and transfer lines, in the assembly category we have fixed position systems, assembly lines and assembly shops (both manual and/or automated operations).

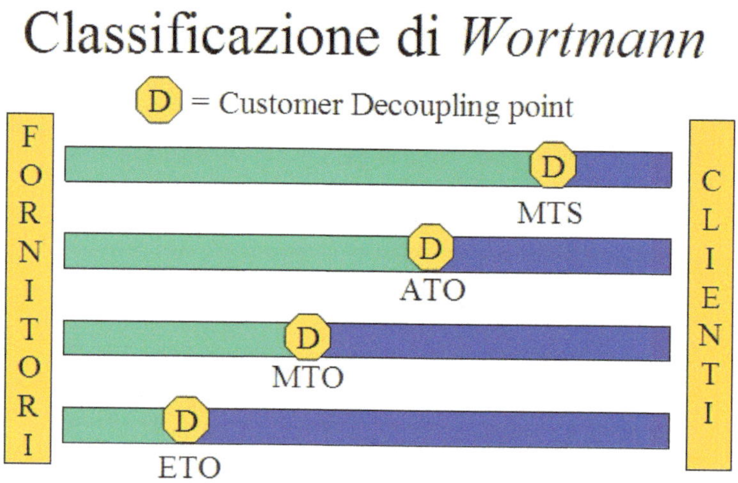

Delivery lead time is the blue bar, manufacturing time is the whole bar, the green bar is the difference between the two.

Another possible classification is one based on Lead Time (manufacturing lead time vs delivery lead time): Engineer to Order (ETO, Purchase to Order (PTO), Make to Order (MTO), Assemble to Order (ATO) and Make to Stock (MTS). According to this classification different kinds of systems will have different customer order decoupling points (CODP), meaning that Work in Progress (WIP) cycle stock levels are practically non-existent regarding operations located after the CODP (except for WIP due to queues).

The concept of production systems can be expanded to the service sector world keeping in mind that services have some fundamental differences in respect to material goods: intangibility, client always present during transformation processes, no stocks for "finished goods". Services can be classified according to a service process matrix: degree of labor intensity (volume) vs degree of customization (variety). With a high degree of labor intensity there are Mass Services (e.g., commercial banking bill payments and state schools) and Professional Services (e.g., personal physicians and lawyers), while with a low degree of labor intensity there are Service Factories (e.g., airlines and hotels) and Service Shops (e.g., hospitals and auto mechanics).

The systems described above are ideal types: real systems may present themselves as hybrids of those categories. Consider, for example, that the production of jeans involves initially carding, spinning, dyeing and weaving, then cutting the fabric in different shapes and assembling the parts in pants or jackets by combining the fabric with thread, zippers and buttons, finally finishing and distressing the pants/jackets before being shipped to stores. The beginning can be seen as process production, the middle as part production and the end again as process production: it's unlikely that a single company will keep all the stages of production under a single roof, therefore the problem of vertical integration and outsourcing arises. Most products require, *from a supply chain perspective*, both process production and part production.

Metrics: Efficiency and Effectiveness

Operations strategy concerns policies and plans of use of the firm productive resources with the aim of supporting long term competitive strategy. Metrics in operations management can be broadly classified into efficiency metrics and effectiveness metrics. Effectiveness metrics involve:

1. Price (actually fixed by marketing, but lower bounded by production cost): purchase price, use costs, maintenance costs, upgrade costs, disposal costs
2. Quality: specification and compliance
3. Time: productive lead time, information lead time, punctuality
4. Flexibility: mix, volume, gamma
5. Stock availability
6. Ecological Soundness: biological and environmental impacts of the system under study.

A more recent approach, introduced by Terry Hill, involves distinguishing competitive variables in order winner and order qualifiers when defining operations strategy. Order winners are variables which permit differentiating the company from competitors, while order qualifiers are prerequisites for engaging in a transaction. This view can be seen as a unifying approach between operations management and marketing.

Productivity is a standard efficiency metric for evaluation of production systems, broadly speaking a ratio between outputs and inputs, and can assume many specific forms, for example: machine productivity, workforce productivity, raw material productivity, warehouse productivity (=inventory turnover). It is also useful to break up productivity in use U (productive percentage of total time) and yield η (ratio between produced volume and productive time) to better evaluate production systems performances. Cycle times can be modeled through manufacturing engineering if the individual operations are heavily automated, if the manual component is the prevalent one, methods used include: time and motion study, predetermined motion time systems and work sampling.

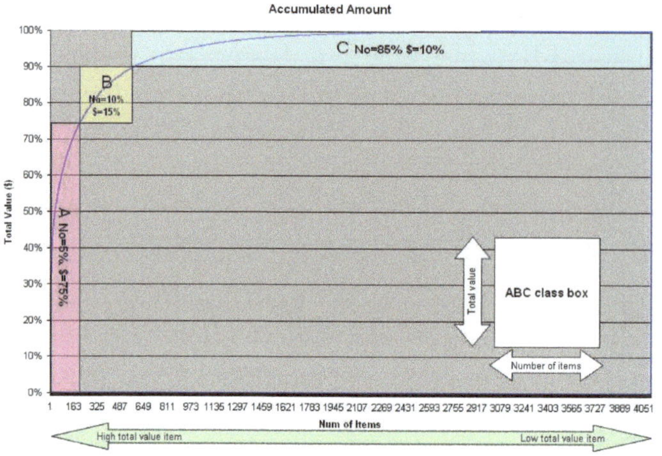

An ABC cumulated curve. Typically one curve is constructed for revenue (consumption) and another for inventory (stock).

ABC analysis is a method for analyzing inventory based on Pareto distribution, it posits that since revenue from items on inventory will be power law distributed then it makes sense to manage items differently based on their position on a revenue-inventory level matrix, 3 classes are constructed (A,B and C) from cumulative item revenues, so in a matrix each item will have a letter (A,B or C) assigned for revenue and inventory. This method posits that items away from the diagonal should be managed differently: items in the upper part are subject to risk of obsolescence, items in the lower part are subject to risk of stockout.

Throughput is a variable which quantifies the number of parts produced in the unit of time. Although estimating throughput for a single process maybe fairly simple, doing so for an entire production system involves an additional difficulty due to the presence of queues which can come from: machine breakdowns, processing time variability, scraps, setups, maintenance time, lack of orders, lack of materials, strikes, bad coordination between resources, mix variability, plus all these inefficiencies tend to compound depending on the nature of the production system. One important example of how system throughput is tied to system design are bottlenecks: in job shops bottlenecks are typically dynamic and dependent on scheduling while on transfer lines it makes sense to speak of "the bottleneck" since it can be univocally associated with a specific station on the line. This leads to the problem of how to define capacity measures, that is an estimation of the maximum output of a given production system, and capacity utilization.

Overall Equipment Effectiveness (OEE) is defined as the product between system availability, cycle time efficiency and quality rate. OEE is typically used as key performance indicator (KPI) in conjunction with the lean manufacturing approach.

Configuration and Management

Designing the *configuration of production systems* involves both technological and organizational variables. Choices in production technology involve: dimensioning capacity, fractioning capacity, capacity location, outsourcing processes, process technology, automation of operations, trade-off between volume and variety. Choices in the organizational area involve: defining worker skills and responsibilities, team coordination, worker incentives and information flow.

Regarding *production planning*, there is a basic distinction between the push approach and the pull approach, with the later including the singular approach of just in time. Pull means that the production system authorizes production based on inventory level; push means that production occurs based on demand (forecasted or present, that is purchase orders). An individual production system can be both push and pull; for example activities before the CODP may work under a pull system, while activities after the CODP may work under a push system.

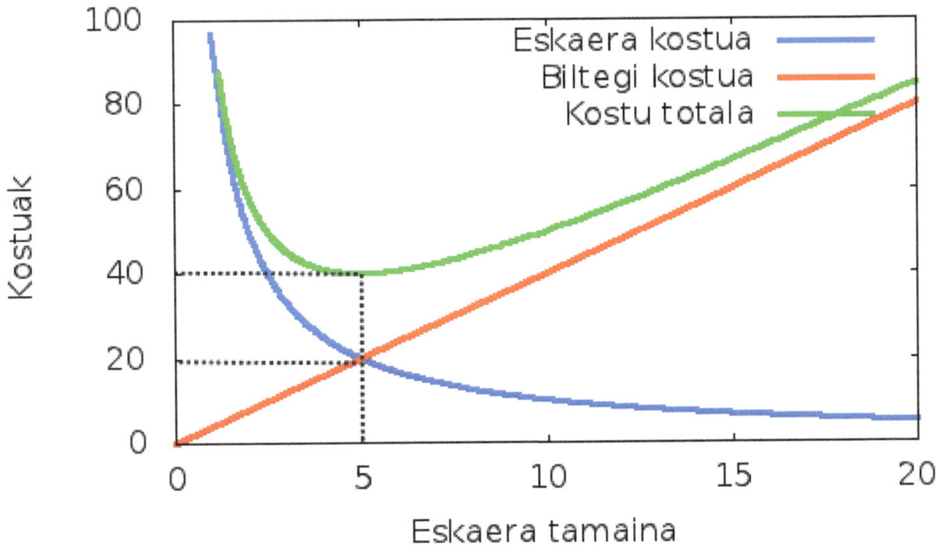

Classic EOQ model: trade-off between ordering cost (blue) and holding cost (red). Total cost (green) admits a global optimum.

Regarding the traditional pull approach to inventory control, a number of techniques have been developed based on the work of Ford W. Harris (1913), which came to be known as the economic order quantity (EOQ) model. This model marks the beginning of inventory theory, which includes the Wagner-Within Procedure, the News Vendor Model, Base Stock Model and the Fixed Time Period model. These models usually involve the calculation of cycle stocks and buffer stocks, the latter usually modeled as a function of demand variability. The Economic Production Quantity (EPQ) differs from the EOQ model only in that it assumes a constant fill rate for the part being produced, instead of the instantaneous refilling of the EOQ model.

Joseph Orlickly and others at IBM developed a push approach to inventory control and production planning, now known as material requirements planning (MRP), which takes as input both the Master Production Schedule (MPS) and the Bill of Materials (BOM) and gives as output a schedule for the materials (components) needed in the production process. MRP therefore is a planning tool to manage purchase orders and production orders (also called jobs).

The MPS can be seen as a kind of aggregate planning for production coming in two fundamentally opposing varieties: plans which try to chase demand and level plans which try to keep uniform capacity utilization. Many models have been proposed to solve MPS problems:

- Analytical models (e.g. Magee Boodman model)
- Exact optimization algorithmic models (e.g. LP and ILP)
- Heuristic models (e.g. Aucamp model).

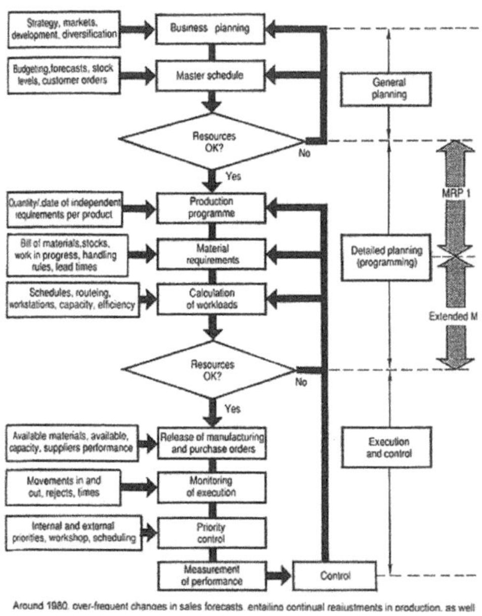

A typical MRPII construct: general planning (top) concerned with forecasts, capacity planning and inventory levels, programming (middle) concerned with calculation of workloads, rough-cut capacity planning, MPS, capacity requirements planning, traditional MRP planning, control (bottom) concerned with scheduling.

MRP can be briefly described as a 3s procedure: sum (different orders), split (in lots), shift (in time according to item lead time). To avoid an "explosion" of data processing in MRP (number of BOMs required in input) planning bills (such as family bills or super bills) can be useful since they allow a rationalization of input data into common codes. MRP had some notorious problems such as infinite capacity and fixed lead times, which influenced successive modifications of the original software architecture in the form of MRP II, enterprise resource planning (ERP) and Advanced planning and scheduling (APS).

In this context problems of scheduling (sequencing of production), loading (tools to use), part type selection (parts to work on) and applications of operations research have a significant role to play.

Lean manufacturing is an approach to production which arose in Toyota between the end of World War II and the seventies. It comes mainly from the ideas of Taiichi Ohno and Toyoda Sakichi which are centered on the complementary notions of just in time and autonomation (jidoka), all aimed at reducing waste (usually applied in PDCA style). Some additional elements are also fundamental: production smoothing (Heijunka), capacity buffers, setup reduction, cross-training and plant layout.

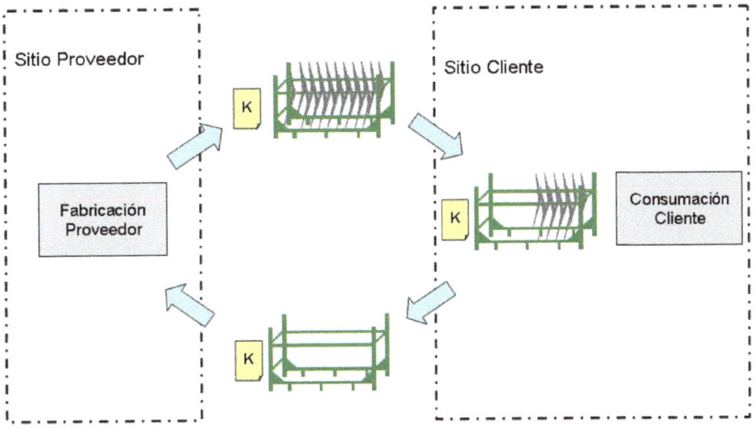

When introducing kanbans in real production systems, attaining unitary lot from the start maybe unfeasible, therefore the kanban will represent a given lot size defined by management.

- Heijunka: production smoothing presupposes a level strategy for the MPS and a final assembly schedule developed from the MPS by smoothing aggregate production requirements in smaller time buckets and sequencing final assembly to achieve repetitive manufacturing. *If these conditions are met*, expected throughput can be equaled to the inverse of takt time. Besides volume, heijunka also means attaining mixed model production, which however may only be feasible through set-up reduction. A standard tool for achieving this is the Heijunka box.
- Capacity buffers: ideally a JIT system would work with zero breakdowns, this however is very hard to achieve in practice, nonetheless Toyota favors acquiring extra capacity over extra WIP to deal with starvation.
- Set-up reduction: typically necessary to achieve mixed model production, a key distinction can be made between internal and external setup. Internal setups (e.g. removing a die) refers to tasks when the machine is not working, while external setups can be completed while the machine is running (ex:transporting dies).
- Cross training: important as an element of Autonomation, Toyota cross trained their employees through rotation, this served as an element of production flexibility, holistic thinking and reducing boredom.

- Layout: U-shaped lines or cells are common in the lean approach since they allow for minimum walking, greater worker efficiency and flexible capacity.

A series of tools have been developed mainly with the objective of replicating Toyota success: a very common implementation involves small cards known as kanbans; these also come in some varieties: reorder kanbans, alarm kanbans, triangular kanbans, etc. In the classic kanban procedure with one card:

- Parts are kept in containers with their respective kanbans
- The downstream station moves the kanban to the upstream station and starts producing the part at the downstream station
- The upstream operator takes the most urgent kanban from his list (compare to queue discipline from queue theory) and produces it and attach its respective kanban

The two-card kanban procedure differs a bit:

- The downstream operator takes the production kanban from his list
- If required parts are available he removes the move kanban and places them in another box, otherwise he chooses another production card
- He produces the part and attach its respective production kanban
- Periodically a mover picks up the move kanbans in upstream stations and search for the respective parts, when found he exchanges production kanbans for move kanbans and move the parts to downstream stations

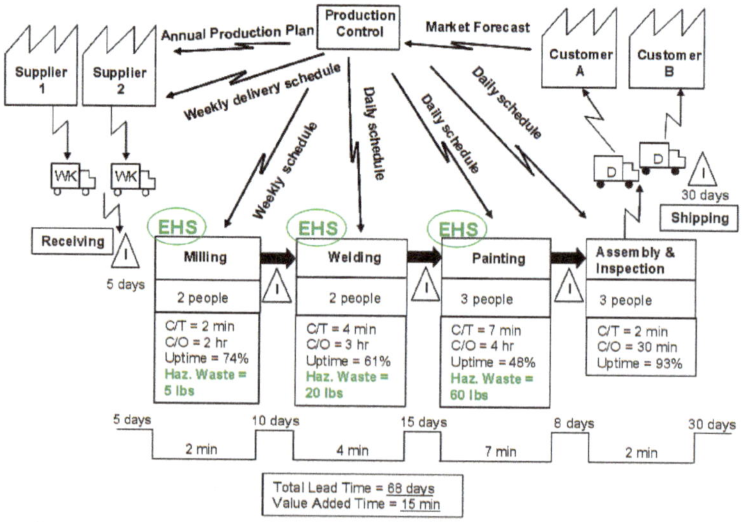

Value Stream Mapping, a representation of materials and information flows inside a company, mainly used in the lean manufacturing approach. The calculation of the time-line (bottom) usually involves using Little's Law to derive lead time from stock levels and takt time.

Since the number of kanbans in the production system is set by managers as a constant number, the kanban procedure works as WIP controlling device, which for a given arrival rate, per Little's Law, works as a lead time controlling device.

In Toyota the TPS represented more of a philosophy of production than a set of specific lean tools, the latter would include:

- SMED: a method for reducing changeover times
- Value Stream Mapping: a graphical method for analyzing the current state and designing a future state
- lot-size reduction
- elimination of time batching
- Rank Order Clustering: an algorithm which groups machines and product families together, used for designing manufacturing cells
- single-point scheduling, the opposite of the traditional push approach
- multi-process handling: when one operator is responsible for operating several machines or processes
- poka-yoke: any mechanism in lean manufacturing that helps an equipment operator avoid (*yokeru*) mistakes (*poka*)
- 5S: describes how to organize a work space for efficiency and effectiveness by identifying and storing the items used, maintaining the area and items, and sustaining the new order
- backflush accounting: a product costing approach in which costing is delayed until goods are finished

Seen more broadly, JIT can include methods such as: product standardization and modularity, group technology, total productive maintenance, job enlargement, job enrichment, flat organization and vendor rating (JIT production is very sensitive to replenishment conditions).

In heavily automated production systems production planning and information gathering may be executed via the control system, attention should be paid however to avoid problems suck as deadlocks, as these can lead to productivity losses.

Mathematical Modeling

There are also fields of mathematical theory which have found applications in the field of operations management such as operations research: mainly mathematical optimization problems and queue theory. Queue theory is employed in modelling queue and processing times in production systems while mathematical optimization draws heavily from multivariate calculus and linear algebra. Queue theory is based on Markov chains and stochastic processes. It also worth noticing that computations of safety stocks are usually based on modeling demand as a normal distribution and MRP and some inventory problems can be formulated using optimal control.

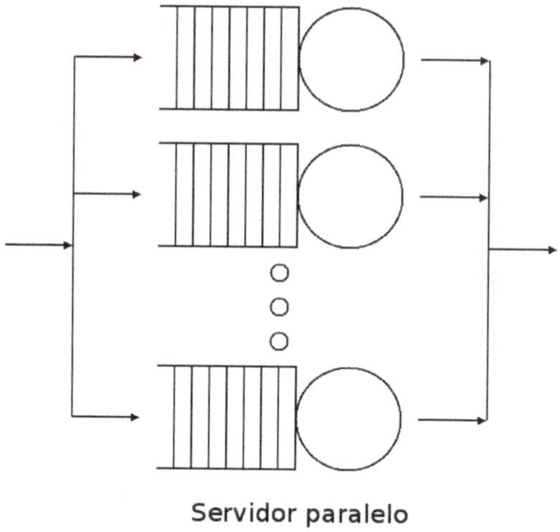

Queue networks are systems in which single queues are connected by a routing network. In this image servers are represented by circles, queues by a series of retangles and the routing network by arrows. In the study of queue networks one typically tries to obtain the equilibrium distribution of the network.

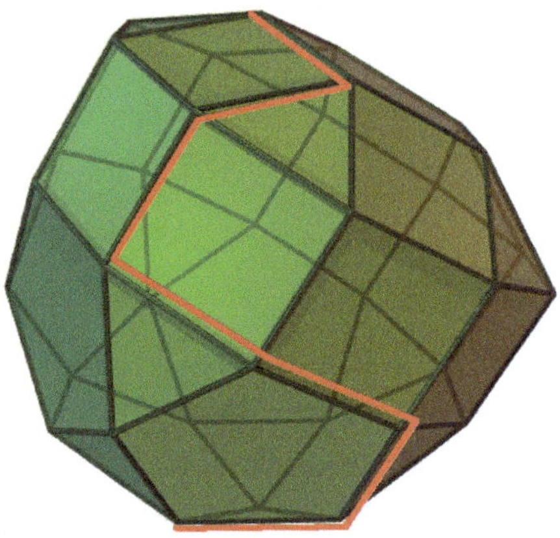

Illustration of the Simplex method, a classical approach to solving LP optimization problems and also integer programming (ex:branch and cut). Mainly used in push approach but also in production system configuration. The interior of the green polytope geometrically represents the feasible region, while the red line indicates the sequence of pivot operations required to reach the optimal solution.

When analytical models are not enough, managers may resort to using simulation. Simulation has been traditionally done thought the Discrete event simulation paradigm, where the simulation model possesses a state which can only change when a discrete event happens, which consists of a clock and list of events. The more recent Transaction-level modeling paradigm consists of a set of resources and a set of transactions: transactions move through a network of resources (nodes) according to a code, called process.

Since real production processes are always affected by disturbances in both inputs and outputs, many companies implement some form of Quality management or quality control. The Seven Basic Tools of Quality designation provides a summary of commonly used tools:

- check sheets
- Pareto charts
- Ishikawa diagrams (Cause-and-effect diagram)
- control charts
- histogram
- scatter diagram
- stratification

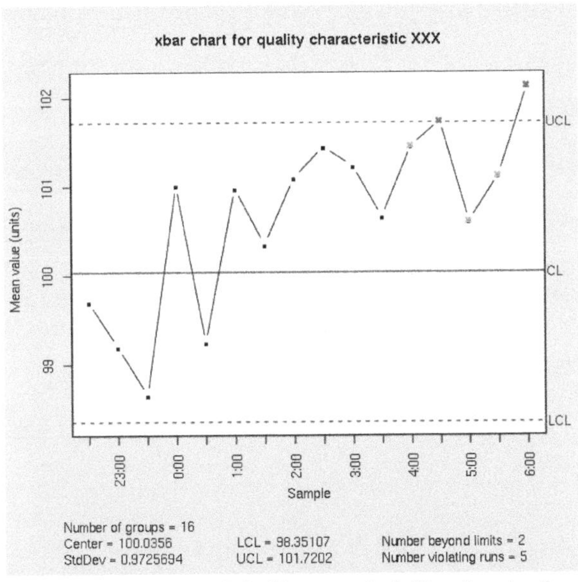

A control chart: process output variable is modeled by a probability density function and for each statistic of the sample an upper control line and lower control line are fixed, when the statistic moves out of bounds, an alarm is given and possible causes are investigated. In this drawing the statistic of choice is the mean and red points represent alarm points.

These are used in approaches like Total quality management and Six Sigma. Keeping quality under control is relevant to both increasing customer satisfaction and reducing processing waste.

Operations management textbooks usually cover demand forecasting, even though it is not strictly speaking an operations problem, because demand is related to some production systems variables. For example, a classic approach in dimensioning safety stocks requires calculating standard deviation of forecast errors. Demand forecasting is also a critical part of push systems, since order releases have to be planned ahead of actual clients orders. Also any serious discussion of capacity planning involves adjusting company outputs with market demands.

Safety, Risk and Maintenance

Other important management problems involve maintenance policies , safety management systems, facility management and supply chain integration.

Organizations

The following organizations support and promote operations management:

- Association for Operations Management (APICS) which supports the *Production and Inventory Management Journal*
- European Operations Management Association (EurOMA) which supports the International Journal of Operations & Production Management
- Production and Operations Management Society (POMS) which supports the journal: *Production and Operations Management*
- Institute for Operations Research and the Management Sciences (INFORMS)
- The Manufacturing and Service Operations Management Society (MSOM) which supports the journal: Manufacturing & Service Operations Management
- Institute of Operations Management (UK)
- Association of Technology, Management, and Applied Engineering (ATMAE)

Material Requirements Planning

Material requirements planning (MRP) is a production planning, scheduling, and inventory control system used to manage manufacturing processes. Most MRP systems are software-based, while it is possible to conduct MRP by hand as well.

An MRP system is intended to simultaneously meet three objectives:

- Ensure materials are available for production and products are available for delivery to customers.
- Maintain the lowest possible material and product levels in store
- Plan manufacturing activities, delivery schedules and purchasing activities.

History

Prior to MRP, and before computers dominated industry, reorder point (ROP)/reorder-quantity (ROQ) type methods like EOQ (economic order quantity) had been used in manufacturing and inventory management.

In 1964, as a response to the Toyota Manufacturing Program, Joseph Orlicky developed material requirements planning (MRP). The first company to use MRP was Black & Decker in 1964, with Dick Alban as project leader. Orlicky's 1975 book *Material Requirements Planning* has the subtitle *The New Way of Life in Production and Inventory Management*. By 1975, MRP was implemented in 700 companies. This number had grown to about 8,000 by 1981.

In 1983, Oliver Wight developed MRP into manufacturing resource planning (MRP II). In the 1980s, Joe Orlicky's MRP evolved into Oliver Wight's manufacturing resource planning (MRP II) which brings master scheduling, rough-cut capacity planning, capacity requirements planning, S&OP in 1983 and other concepts to classical MRP. By 1989, about one third of the software industry was MRP II software sold to American industry ($1.2 billion worth of software).

The Scope of MRP in Manufacturing

Dependent Demand vs Independent Demand

Independent demand is demand originating outside the plant or production system, while dependent demand is demand for components. The bill of materials (BOM) specifies the relationship between the end product (independent demand) and the components (dependent demand). MRP take as input the information contained in the BOM.

The basic functions of an MRP system include: inventory control, bill of material processing, and elementary scheduling. MRP helps organizations to maintain low inventory levels. It is used to plan manufacturing, purchasing and delivering activities.

"Manufacturing organizations, whatever their products, face the same daily practical problem - that customers want products to be available in a shorter time than it takes to make them. This means that some level of planning is required."

Companies need to control the types and quantities of materials they purchase, plan which products are to be produced and in what quantities and ensure that they are able to meet current and future customer demand, all at the lowest possible cost. Making a bad decision in any of these areas will make the company lose money. A few examples are given below:

- If a company purchases insufficient quantities of an item used in manufacturing (or the wrong item) it may be unable to meet contract obligations to supply products on time.

- If a company purchases excessive quantities of an item, money is wasted - the excess quantity ties up cash while it remains as stock and may never even be used at all.
- Beginning production of an order at the wrong time can cause customer deadlines to be missed.

MRP is a tool to deal with these problems. It provides answers for several questions:

- *What* items are required?
- *How many* are required?
- *When* are they required?...

MRP can be applied both to items that are purchased from outside suppliers and to sub-assemblies, produced internally, that are components of more complex items.

Data

The data that must be considered include:

- The *end item* (or items) being created. This is sometimes called independent demand, or Level "0" on BOM (bill of materials).
- How much is required at a time.
- When the quantities are required to meet demand.
- Shelf life of stored materials.
- Inventory status records. Records of *net* materials *available* for use already in stock (on hand) and materials on order from suppliers.
- Bills of materials. Details of the materials, components and sub-assemblies required to make each product.
- Planning data. This includes all the restraints and directions to produce the end items. This includes such items as: routing, labor and machine standards, quality and testing standards, pull/work cell and push commands, lot sizing techniques (i.e. fixed lot size, lot-for-lot, economic order quantity), scrap percentages, and other inputs.

Outputs

There are two outputs and a variety of messages/reports:

- Output 1 is the "Recommended Production Schedule" which lays out a detailed schedule of the required minimum start and completion dates, with quantities, for each step of the Routing and Bill Of Material required to satisfy the demand from the master production schedule (MPS).
- Output 2 is the "Recommended Purchasing Schedule". This lays out both the dates that the purchased items should be received into the facility *and* the dates that the purchase orders or blanket order release should occur to match the production schedules.

Messages and reports:

- Purchase orders. An order to a supplier to provide materials.
- Reschedule notices. These *recommend* cancelling, increasing, delaying or speeding up existing orders.

Methods to Find Order Quantities

Well-known methods to find order quantities are:

- Dynamic lot-sizing
- Silver–Meal heuristic
- Least-Unit-Cost heuristic

Mathematical Formulation

Initial conditions:

$$x_i'(0) = x_{i0}' i = 1,\dots,J$$

Dynamics:

$$x_i'(t+1) = x_i'(t) + z_i(t) - d_i'(t) - \sum_{j \in Suc(i)} z_j(t + L_{ij}' - 1)$$

t=0,...,T-1,i=1,...,J

Constraints:

$$x_i'(t) \geq 0 t \text{ t=1,...,T,i=1,...,J}$$

$$z_i(t) \geq 0 \text{ t=0,...,T-1,i=1,...,J}$$

Objective:

$$min \sum_i \sum_{t=0}^{T-1} [k_i(t)\delta(z_i(t)) + c_i(t) z_i(t)] + \sum_i \sum_{t=1}^{T} h_i'(t) x_i'(t)$$

MRP can be expressed as an optimal control problem:

Where x' is local inventory (the state), z the order size (the control), d is local demand, k represents fixed order costs, c variable order costs, h local inventory holding costs. $\delta()$ is the Heaviside function. Changing the dynamics of the problem leads to a multi-item analogue of the dynamic lot-size model.

Problems with MRP Systems

- Integrity of the data. If there are any errors in the inventory data, the bill of materials (commonly referred to as 'BOM') data, or the master production schedule, then the output data will also be incorrect ("GIGO": garbage in, garbage out). Data integrity is also affected by inaccurate cycle count adjustments, mistakes in receiving input and shipping output, scrap not reported, waste, damage, box count errors, supplier container count errors, production reporting errors, and system issues. Many of these type of errors can be minimized by implementing pull systems and using bar code scanning. Most vendors in this type of system recommend at least 99% data integrity for the system to give useful results.
- Systems require that the user specify how long it will take for a factory to make a product from its component parts (assuming they are all available). Additionally, the system design also assumes that this "lead time" in manufacturing will be the same each time the item is made, without regard to quantity being made, or other items being made simultaneously in the factory.
- A manufacturer may have factories in different cities or even countries. It is not good for an MRP system to say that we do not need to order some material, because we have plenty of it thousands of miles away. The overall ERP system needs to be able to organize inventory and needs by individual factory and inter-communicate the needs in order to enable each factory to redistribute components to serve the overall enterprise. This means that other systems in the enterprise need to work properly, both before implementing an MRP system and in the future. For example, systems like variety reduction and engineering, which makes sure that product comes out right first time (without defects), must be in place.
- Production may be in progress for some part, whose design gets changed, with customer orders in the system for both the old design, and the new one, concurrently. The overall ERP system needs to have a system of coding parts such that the MRP will correctly calculate needs and tracking for both versions. Parts must be booked into and out of stores more regularly than the MRP calculations take place. Note, these other systems can well be manual systems, but must interface to the MRP. For example, a 'walk around' stock intake done just prior to the MRP calculations can be a practical solution for a small inventory (especially if it is an "open store").

- The other major drawback of MRP is that it fails to account for capacity in its calculations. This means it will give results that are impossible to implement due to manpower, machine or supplier capacity constraints. However this is largely dealt with by MRP II. Generally, MRP II refers to a system with integrated financials. An MRP II system can include finite or infinite capacity planning. But, to be considered a true MRP II system must also include financials. In the MRP II (or MRP2) concept, fluctuations in forecast data are taken into account by including simulation of the master production schedule, thus creating a long-term control. A more general feature of MRP2 is its extension to purchasing, to marketing and to finance (integration of all the functions of the company), ERP has been the next step.

Solutions to Data Integrity Issues

- Bill of material – The best practice is to physically verify the bill of material either at the production site or by disassembling the product.
- Cycle count – The best practice is to determine why a cycle count that increases or decreases inventory has occurred. Find the root cause and correct the problem from occurring again.
- Scrap reporting – This can be the most difficult area to maintain with any integrity. Start with isolating the scrap by providing scrap bins at the production site and then record the scrap from the bins on a daily basis. One benefit of reviewing the scrap on site is that preventive action can be taken by the engineering group.
- Receiving errors – Manual systems of recording what has been received are error prone. The best practice is to implement the system of receiving by ASN from the supplier. The supplier sends an ASN (advanced shipping notification). When the components are received into the facility, the ASN is processed and then company labels are created for each line item. The labels are affixed to each container and then scanned into the MRP system. Extra labels reveal a shortage from the shipment and too few labels reveal an over shipment. Some companies pay for ASN by reducing the time in processing accounts payable.
- Shipping errors – The container labels are printed from the shipper. The labels are affixed to the containers in a staging area or when they are loaded on the transport.
- Production reporting – The best practice is to use bar code scanning to enter production into inventory. A product that is rejected should be moved to an MRB (material review board) location. Containers that require sorting need to be received in reverse.

- Replenishment – The best replenishment practice is replacement using bar code scanning, or via pull system. Depending upon the complexity of the product, planners can actually order materials using scanning with a min-max system.

Demand Driven MRP

In 2011, the third edition of "Orlicky's Planning" introduced a new type of MRP called "demand driven MRP" (DDMRP). The new edition of the book was written, not by Orlicky himself (he died in 1986) but by Carol Ptak and Chad Smith at the invitation of McGraw Hill to update Orlicky's work.

Demand driven MRP is a multi-echelon formal planning and execution technique with five distinct components:

1. Strategic inventory positioning – The first question of effective inventory management is not, "how much inventory should we have?" Nor is it, "when should we make or buy something?" The most fundamental question to ask in today's manufacturing environments is, "given our system and environment, where should we place inventory to have the best protection?" Inventory is like a break wall to protect boats in a marina from the roughness of incoming waves. Out on the open ocean the break walls have to be 50–100 feet tall, but in a small lake the break walls are only a couple feet tall. In a glassy smooth pond no break wall is necessary.

2. Buffer profiles and level – Once the strategically replenished positions are determined, the actual levels of those buffers have to be initially set. Based on several factors, different materials and parts behave differently (but many also behave nearly the same). DDMRP calls for the grouping of parts and materials chosen for strategic replenishment and that behave similarly into "buffer profiles." Buffer profiles take into account important factors including lead time (relative to the environment), variability (demand or supply), whether the part is made or bought or distributed and whether there are significant order multiples involved. These buffer profiles are made up of "zones" that produce a unique buffer picture for each part as their respective individual part traits are applied to the group traits.

3. Dynamic adjustments – Over the course of time, group and individual traits can and will change as new suppliers and materials are used, new markets are opened and/or old markets deteriorate and manufacturing capacities and methods change. Dynamic buffer levels allow the company to adapt buffers to group and individual part trait changes over time through the use of several types of adjustments. Thus, as more or less variability is encountered or as a company's strategy changes these buffers adapt and change to fit the environment.

4. Demand-driven planning – takes advantage of the sheer computational power of today's hardware and software. It also takes advantage of the new de-

mand-driven or pull-based approaches. When these two elements are combined then there is the best of both worlds; relevant approaches and tools for the way the world works today and a system of routine that promotes better and quicker decisions and actions at the planning and execution level.

5. Highly visible and collaborative execution – Simply launching purchase orders (POs), manufacturing orders (MOs) and transfer orders (TOs) from any planning system does not end the materials and order management challenge. These POs, MOs and TOs have to be effectively managed to synchronize with the changes that often occur within the "execution horizon." The execution horizon is the time from which a PO, MO or TO is opened until the time it is closed in the system of record. DDMRP defines a modern, integrated and greatly needed system of execution for all part categories in order to speed the proliferation of relevant information and priorities throughout an organization and supply chain.

These five components work together to greatly dampen, if not eliminate, the nervous-ness of traditional MRP systems and the bullwhip effect in complex and challenging environments. Many claim have been made by the consultancy company that is mar-keting DDMRP, including the following: In utilizing these approaches, planners will no longer have to try to respond to every single message for every single part that is off by even one day. This approach provides real information about those parts that are truly at risk of negatively impacting the planned availability of inventory. DDMRP sorts the significant few items that require attention from the many parts that are being man-aged. Under the DDMRP approach, fewer planners can make better decisions more quickly. That means companies will be better able to leverage their working and hu-man capital as well as the huge investments they have made in information technology. One down-side, however, is that DDMRP can not run on the majority of MRPII/ERP systems in use today, so companies that wish to use it have to augment their current system with a 'certified compliant' system.

DDMRP has been successfully applied to a variety of environments including CTO (configure to order), MTS (make to stock), MTO (make to order) and ETO (engineer to order). The methodology is applied differently in each environments but the five step process remains the same. DDMRP leverages knowledge from theory of constraints (TOC), traditional MRP & DRP, Six Sigma and lean. It is effectively an amalgam of MRP and kanban techniques.

Process Layout

In manufacturing engineering, process layout is a design for the floor plan of a plant which aims to improve efficiency by arranging equipment according to its function. The production line should ideally be designed to eliminate waste in material flows, inventory handling and management. In process layout, the work stations and machinery are not arranged according to a particular production sequence. Instead, there is an assembly of similar operations or similar machinery in each department (for example, a drill department, a paint department, etc.)

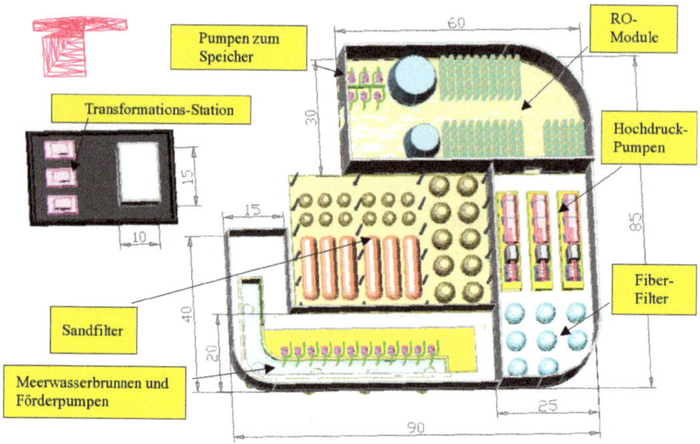

Typical plant layout done thought CAD. Note that layout for a production system reflects a conceived organization of resources to achieve a certain goal, besides satisfying certain space constraints.

Main Advantages

1. Provide visual control of activities
2. Utilize space efficiently
3. Utilize labor efficiently
4. Eliminate bottlenecks
5. Facilitate communication and interaction between workers and supervisors

Criticism

A common criticism of this layout is that the work can be monotonous for staff, especially if they are involved only in one stage of the process. This criticism can however be eliminated if the staff is rotated to different departments (involving different processes) thus developing a multi-skilled staff.

Production Line

A production line is a set of sequential operations established in a factory whereby materials are put through a refining process to produce an end-product that is suitable for onward consumption; or components are assembled to make a finished article.

Typically, raw materials such as metal ores or agricultural products such as foodstuffs or textile source plants (cotton, flax) require a sequence of treatments to render them useful. For metal, the processes include crushing, smelting and further refining. For plants, the useful material has to be separated from husks or contaminants and then treated for onward sale.

Doughnut production line

History

Early production processes were constrained by the availability of a source of energy, with wind mills and water mills providing power for the crude heavy processes and manpower being used for activities requiring more precision. In earlier centuries, with raw materials, power and people often being in different locations, production was distributed across a number of sites. The concentration of numbers of people in manufactories, and later the factory as exemplified by the cotton mills of Richard Arkwright, started the move towards co-locating individual processes.

Introduction of the Steam Engine

With the development of the steam engine in the latter half of the 18th century, the production elements became less reliant on the location of the power source, and so the processing of goods moved to either the source of the materials or the location of people to perform the tasks. Separate processes for different treatment stages were brought into the same building, and the various stages of refining or manufacture were combined.

Industrial Revolution

With increasing use of steam power, and increasing use of machinery to supplant the use of people, the integrated use of techniques in production lines spurred the industrial revolutions of Europe and North America.

Assembly Line

Thus, from the processing of raw materials into useful goods, the next step was the concept of the assembly line, as introduced by Eli Whitney. This was taken to the next stage at the Ford Motor Company in 1913, where Henry Ford introduced the innovation of continuously moving the cars being assembled past individual work stations. This introduced the idea of standardization.

The assembly line was a contraption of many chains and links that moved to place different parts into various places throughout the car. The chassis of the car was along the 45 metre line by a chain conveyor and then 140 workers applied their assigned parts to the chassis. Other workers brought additional parts to the car builders to keep them stocked. The assembly line decreased the assembly time per vehicle. The production time for a single car dropped from over twelve hours to just 93 minutes.

Assembly Line

An assembly line is a manufacturing process (most of the time called a progressive assembly) in which parts (usually interchangeable parts) are added as the semi-finished assembly moves from workstation to workstation where the parts are added in sequence until the final assembly is produced. By mechanically moving the parts to the assembly work and moving the semi-finished assembly from work station to work station, a finished product can be assembled faster and with less labor than by having workers carry parts to a stationary piece for assembly.

An Airbus A321 on final assembly line 3 in the Airbus plant at Hamburg Finkenwerder Airport.

Assembly lines are common methods of assembling complex items such as automobiles and other transportation equipment, household appliances and electronic goods.

Hyundai's car assembly line.

Concepts

Assembly lines are designed for the sequential organization of workers, tools or machines, and parts. The motion of workers is minimized to the extent possible. All parts or assemblies are handled either by conveyors or motorized vehicles such as fork lifts, or gravity, with no manual trucking. Heavy lifting is done by machines such as overhead cranes or fork lifts. Each worker typically performs one simple operation.

Lotus Cars assembly line as of 2008

According to Henry Ford:

The principles of assembly are these:

(1) Place the tools and the men in the sequence of the operation so that each component part shall travel the least possible distance while in the process of finishing.

(2) Use work slides or some other form of carrier so that when a workman completes his operation, he drops the part always in the same place—which place must always be the most convenient place to his hand—and if possible have gravity carry the part to the next workman for his own.

(3) Use sliding assembling lines by which the parts to be assembled are delivered at convenient distances.

Simple Example

Consider the assembly of a car: assume that certain steps in the assembly line are to install the engine, install the hood, and install the wheels (in that order, with arbitrary interstitial steps); only one of these steps can be done at a time. In traditional production, only one car would be assembled at a time. If engine installation takes 20 minutes, hood installation takes five minutes, and wheels installation takes 10 minutes, then a car can be produced every 35 minutes.

In an assembly line, car assembly is split between several stations, all working simultaneously. When one station is finished with a car, it passes it on to the next. By having three stations, a total of three different cars can be operated on at the same time, each one at a different stage of its assembly.

After finishing its work on the first car, the engine installation crew can begin working on the second car. While the engine installation crew works on the second car, the first car can be moved to the hood station and fitted with a hood, then to the wheels station and be fitted with wheels. After the engine has been installed on the second car, the second car moves to the hood assembly. At the same time, the third car moves to the engine assembly. When the third car's engine has been mounted, it then can be moved to the hood station; meanwhile, subsequent cars (if any) can be moved to the engine installation station.

Assuming no loss of time when moving a car from one station to another, the longest stage on the assembly line determines the throughput (20 minutes for the engine installation) so a car can be produced every 20 minutes, once the first car taking 35 minutes has been produced.

History

Before the Industrial Revolution, most manufactured products were made individually by hand. A single craftsman or team of craftsmen would create each part of a product. They would use their skills and tools such as files and knives to create the individual parts. They would then assemble them into the final product, making cut-and-try changes in the parts until they fit and could work together (craft production).

Division of labor was practiced in China where state run monopolies mass-produced metal agricultural implements, china, armor,and weapons centuries before it appeared in Europe on the eve of the Industrial Revolution. Adam Smith discussed the division of labour in the manufacture of pins at length in his book *The Wealth of Nations* (published in 1776).

The Venetian Arsenal, dating to about 1104, operated similar to a production line. Ships moved down a canal and were fitted by the various shops they passed. At the peak of its efficiency in the early 16th century, the Venetian Arsenal employed some 16,000 people who could apparently produce nearly one ship each day, and could fit out, arm, and provision a newly built galley with standardized parts on an assembly-line basis. Although the Venice Arsenal lasted until the early Industrial Revolution, production line methods did not become common even then.

Industrial Revolution

The pulley block was the first manufacture to become fully automated at the Portsmouth Block Mills in the early 19th century.

The Industrial Revolution led to a proliferation of manufacturing and invention. Many industries, notably textiles, firearms, clocks and watches, horse-drawn vehicles, railway locomotives, sewing machines, and bicycles, saw expeditious improvement in materials handling, machining, and assembly during the 19th century, although modern concepts such as industrial engineering and logistics had not yet been named.

The automatic flour mill built by Oliver Evans in 1785 was called the beginning of modern bulk material handling by Roe (1916). Evans's mill used a leather belt bucket elevator, screw conveyors, canvas belt conveyors, and other mechanical devices to completely automate the process of making flour. The innovation spread to other mills and breweries.

Probably the earliest industrial example of a linear and continuous assembly process is the Portsmouth Block Mills, built between 1801 and 1803. Marc Isambard Brunel (father of Isambard Kingdom Brunel), with the help of Henry Maudslay and others, designed 22 types of machine tools to make the parts for the rigging blocks used by the Royal Navy. This factory was so successful that it remained in use until the 1960s, with the workshop still visible at HM Dockyard in Portsmouth, and still containing some of the original machinery.

One of the earliest examples of an almost modern factory layout, designed for easy material handling, was the Bridgewater Foundry. The factory grounds were bordered by the Bridgewater Canal and the Liverpool and Manchester Railway. The buildings were arranged in a line with a railway for carrying the work going through the buildings. Cranes were used for lifting the heavy work, which sometimes weighed in the tens of tons. The work passed sequentially through to erection of framework and final assembly.

The Bridgewater Foundry, pictured in 1839, one of the earliest factories to use an almost modern layout, workflow, and material-handling system.

The first flow assembly line was initiated at the factory of Richard Garrett & Sons, Leiston Works in Leiston in the English county of Suffolk for the manufacture of portable steam engines. The assembly line area was called 'The Long Shop' on account of its length and was fully operational by early 1853. The boiler was brought up from the foundry and put at the start of the line, and as it progressed through the building it would stop at various stages where new parts would be added. From the upper level, where other parts were made, the lighter parts would be lowered over a balcony and then fixed onto the machine on the ground level. When the machine reached the end of the shop, it would be completed.

Interchangeable Parts

During the early 19th century, the development of machine tools such as the screw-cutting lathe, metal planer, and milling machine, and of toolpath control via jigs and fixtures, provided the prerequisites for the modern assembly line by making interchangeable parts a practical reality.

Late 19th Century Steam and Electric Conveyors

Steam powered conveyor lifts began being used for loading and unloading ships some time in the last quarter of the 19th century. Hounshell (1984) shows a c. 1885 sketch of an electric powered conveyor moving cans through a filling line in a canning factory.

The meatpacking industry of Chicago is believed to be one of the first industrial assembly lines (or dis-assembly lines) to be utilized in the United States starting in 1867. Workers would stand at fixed stations and a pulley system would bring the meat to each worker and they would complete one task. Henry Ford and others have written about the influence of this slaughterhouse practice on the later developments at Ford Motor Company.

20th Century

Ford assembly line, 1913. The magneto assembly line was the first.

1913 Experimenting with mounting body on Model T chassis. Ford tested various as-sembly methods to optimize the procedures before permanently installing the equip-ment. The actual assembly line used an overhead crane to mount the body.

The moving assembly line was developed for the Ford Model T and began operation on

October 7, 1913. The assembly line, driven by conveyor belts, reduced production time for a Model T to just 93 minutes. Producing cars quicker than paint of the day could dry, it had an immense influence on the world.

According to Domm, the implementation of mass production of an automobile via an assembly line may be credited to Ransom Olds, who used it to build the first mass-produced automobile, the Oldsmobile Curved Dash. Olds patented the assembly line concept, which he put to work in his Olds Motor Vehicle Company factory in 1901. Henry Ford,

At Ford Motor Company, the assembly line concept appears to have been introduced by William "Pa" Klann upon his return from visiting Swift & Company's slaughterhouse in Chicago and viewing what was referred to as the "disassembly line", where carcasses were butchered as they moved along a conveyor. The efficiency of one person removing the same piece over and over without himself moving caught his attention. He reported the idea to Peter E. Martin, soon to be head of Ford production, who was doubtful at the time but encouraged him to proceed. Others at Ford have claimed to have put the idea forth to Henry Ford, but Pa Klann's slaughterhouse revelation is well documented in the archives at the Henry Ford Museum and elsewhere, making him an important contributor to the modern automated assembly line concept. The process was an evolution by trial and error of a team consisting primarily of Peter E. Martin, the factory superintendent; Charles E. Sorensen, Martin's assistant; C. Harold Wills, draftsman and toolmaker; Clarence W. Avery; Charles Ebender; and József Galamb. Some of the groundwork for such development had recently been laid by the intelligent layout of machine tool placement that Walter Flanders had been doing at Ford up to 1908.

In 1922 Ford (through his ghostwriter Crowther) said of his 1913 assembly line:

I believe that this was the first moving line ever installed. The idea came in a general way from the overhead trolley that the Chicago packers use in dressing beef.

Charles E. Sorensen, in his 1956 memoir *My Forty Years with Ford*, presented a different version of development that was not so much about individual "inventors" as a gradual, logical development of industrial engineering:

What was worked out at Ford was the practice of moving the work from one worker to another until it became a complete unit, then arranging the flow of these units at the right time and the right place to a moving final assembly line from which came a finished product. Regardless of earlier uses of some of these principles, the direct line of succession of mass production and its intensification into automation stems directly from what we worked out at Ford Motor Company between 1908 and 1913. Henry Ford is generally regarded as the father of mass production. He was not. He was the sponsor of it.

As a result of these developments in method, Ford's cars came off the line in three-minute intervals. This was much faster than previous methods, increasing production by eight to one (requiring 12.5 man-hours before, 1 hour 33 minutes after), while using less manpower. It was so successful, paint became a bottleneck. Only japan black would dry fast enough, forcing the company to drop the variety of colors available before 1914, until fast-drying Duco lacquer was developed in 1926.

The assembly line technique was an integral part of the diffusion of the automobile into American society. Decreased costs of production allowed the cost of the Model T to fall within the budget of the American middle class. In 1908, the price of a Model T was around $825, and by 1912 it had decreased to around $575. This price reduction is comparable to a reduction from $15,000 to $10,000 in dollar terms from the year 2000. In 1914, an assembly line worker could buy a Model T with four months' pay.

Ford's complex safety procedures—especially assigning each worker to a specific location instead of allowing them to roam about—dramatically reduced the rate of injury. The combination of high wages and high efficiency is called "Fordism", and was copied by most major industries. The efficiency gains from the assembly line also coincided with the take-off of the United States. The assembly line forced workers to work at a certain pace with very repetitive motions which led to more output per worker while other countries were using less productive methods.

In the automotive industry, its success was dominating, and quickly spread worldwide. Ford France and Ford Britain in 1911, Ford Denmark 1923, Ford Germany 1925; in 1919, Vulcan (Southport, Lancashire) was the first native European manufacturer to adopt it. Soon, companies had to have assembly lines, or risk going broke by not being able to compete; by 1930, 250 companies which did not had disappeared.

The massive demand for military hardware in World War II prompted assembly-line techniques in shipbuilding and aircraft production. Thousands of Liberty Ships were built making extensive use of prefabrication, enabling ship assembly to be completed in weeks or even days. After having produced fewer than 3,000 planes for the United States Military in 1939, American aircraft manufacturers built over 300,000 planes in World War II. Vultee pioneered the use of the powered assembly line for aircraft manufacturing. Other companies quickly followed. As William S. Knudsen of the National Defense Advisory Commission observed, "We won because we smothered the enemy in an avalanche of production, the like of which he had never seen, nor dreamed possible."

Improved Working Conditions

In his 1922 autobiography, Henry Ford mentions several benefits of the assembly line including:

- Workers do no heavy lifting.
- No stooping or bending over.

- No special training required.
- There are jobs that almost anyone can do.
- Provided employment to immigrants.

The gains in productivity allowed Ford to increase worker pay from $1.50 per day to $5.00 per day once employees reached three years of service on the assembly line. Ford continued on to reduce the hourly work week while continuously lowering the Model T price. These goals appear altruistic; however, it has been argued that they were implemented by Ford in order to reduce high employee turnover: when the assembly line was introduced in 1913, it was discovered that "every time the company wanted to add 100 men to its factory personnel, it was necessary to hire 963" in order to counteract the natural distaste the assembly line seems to have inspired.

Sociological Problems

Sociological work has explored the social alienation and boredom that many workers feel because of the repetition of doing the same specialized task all day long.

One of Capitalism's most famous critics: Karl Marx highlighted in the *Entfremdung* theory the fact that in order to achieve job satisfaction workers need to see themselves in the objects they've created, products should be "mirrors in which workers see their reflected essential nature." Marx viewed labour as a chance for us to externalize facets of our personality. They argue that specialization makes it incredibly difficult for any worker to achieve a sense of genuine contribution they may be making to the real needs of humanity. The repetitive nature of specialized tasks causes, they say, a feeling of disconnection between what a worker does all day and who they really are and would ideally be able to contribute to existence. Marx also argued that specialised jobs are insecure, since they make the worker utterly expendable the minute costs rise and savings can be made through technology.

Since workers have to stand in the same place for hours and repeat the same motion hundreds of times per day repetitive stress injuries are a possible pathology of occupational safety. Industrial noise also proved dangerous. When it was not too high, workers were often prohibited from talking. Charles Piaget, a skilled worker at the LIP factory, recalled that beside being prohibited from speaking, the semi-skilled workers had only 25 centimeters in which to move. Industrial ergonomics later tried to minimize physical trauma.

Prefabrication

Prefabrication is the practice of assembling components of a structure in a factory or other manufacturing site, and transporting complete assemblies or sub-assemblies to the construction site where the structure is to be located. The term is used to distinguish this process from the more conventional construction practice of transporting the basic materials to the construction site where all assembly is carried out.

The term prefabrication also applies to the manufacturing of things other than structures at a fixed site. It is frequently used when fabrication of a section of a machine or any movable structure is shifted from the main manufacturing site to another location, and the section is supplied assembled and ready to fit. It is not generally used to refer to electrical or electronic components of a machine, or mechanical parts such as pumps, gearboxes and compressors which are usually supplied as separate items, but to sections of the body of the machine which in the past were fabricated with the whole machine. Prefabricated parts of the body of the machine may be called 'sub-assemblies' to distinguish them from the other components.

Process and Theory

An example from house-building illustrates the process of prefabrication. The conventional method of building a house is to transport bricks, timber, cement, sand, steel and construction aggregate, etc. to the site, and to construct the house on site from these materials. In prefabricated construction, only the foundations are constructed in this way, while sections of walls, floors and roof are prefabricated (assembled) in a factory (possibly with window and door frames included), transported to the site, lifted into place by a crane and bolted together.

Prefabrication is used in the manufacture of ships, aircraft and all kinds of vehicles and machines where sections previously assembled at the final point of manufacture are assembled elsewhere instead, before being delivered for final assembly.

The theory behind the method is that time and cost is saved if similar construction tasks can be grouped, and assembly line techniques can be employed in prefabrication at a location where skilled labour is available, while congestion at the assembly site, which wastes time, can be reduced. The method finds application particularly where the structure is composed of repeating units or forms, or where multiple copies of the same basic structure are being constructed. Prefabrication avoids the need to transport so many skilled workers to the construction site, and other restricting conditions such as a lack of power, lack of water, exposure to harsh weather or a hazardous environment are avoided. Against these advantages must be weighed the cost of transporting prefabricated sections and lifting them into position as they will usually be larger, more fragile and more difficult to handle than the materials and components of which they are made.

History

Prefabrication has been used since ancient times. For example, it is claimed that the world's oldest known engineered roadway, the Sweet Track constructed in England around 3800 BC, employed prefabricated timber sections brought to the site rather than assembled on-site.

"Loren" Iron House, at Old Gippstown in Moe, Australia

Sinhalese kings of ancient Sri Lanka have used prefabricated buildings technology to erect giant structures, which dates back as far as 2000 years, where some sections were prepared separately and then fitted together, specially in the Kingdom of Anuradhapura and Kingdom of Polonnaruwa

In 19th century Australia a large number of prefabricated houses were imported from the United Kingdom.

The method was widely used in the construction of prefabricated housing in the 20th century, such as in the United Kingdom as temporary housing for thousands of urban families "bombed out" during World War II. Assembling sections in factories saved time on-site and the lightness of the panels reduced the cost of foundations and assembly on site. Coloured concrete grey and with flat roofs, prefab houses were uninsulated and cold and life in a prefab acquired a certain stigma, but some London prefabs were occupied for much longer than the projected 10 years.

The Crystal Palace, erected in London in 1851, was a highly visible example of iron and glass prefabricated construction; it was followed on a smaller scale by Oxford Rewley Road railway station.

Current uses

The most widely used form of prefabrication in building and civil engineering is the use of prefabricated concrete and prefabricated steel sections in structures where a particular part or form is repeated many times. It can be difficult to construct the formwork required to mould concrete components on site, and delivering wet concrete to the site before it starts to set requires precise time management. Pouring concrete sections in a factory brings the advantages of being able to re-use moulds and the concrete can be mixed on the spot without having to be transported to and pumped wet on a congested construction site. Prefabricating steel sections reduces on-site cutting and welding costs as well as the associated hazards.

A house being built with prefabricated concrete panels.

Prefabrication techniques are used in the construction of apartment blocks, and housing developments with repeated housing units. The quality of prefabricated housing units had increased to the point that they may not be distinguishable from traditionally built units to those that live in them. The technique is also used in office blocks, warehouses and factory buildings. Prefabricated steel and glass sections are widely used for the exterior of large buildings.

Detached houses, cottages, log cabin, saunas, etc. are also sold with prefabricated elements. Prefabrication of modular wall elements allows building of complex thermal insulation, window frame components, etc. on an assembly line, which tends to improve quality over on-site construction of each individual wall or frame. Wood construction in particular benefits from the improved quality. However, tradition often favors building by hand in many countries, and the image of prefab as a "cheap" method only slows its adoption. However, current practice already allows the modifying the floor plan according to the customer's requirements and selecting the surfacing material, e.g. a personalized brick facade can be masoned even if the load-supporting elements are timber.

Prefabrication saves engineering time on the construction site in civil engineering projects. This can be vital to the success of projects such as bridges and avalanche galleries, where weather conditions may only allow brief periods of construction. Prefabricated bridge elements and systems offer bridge designers and contractors significant advantages in terms of construction time, safety, environmental impact, constructibility, and cost. Prefabrication can also help minimize the impact on traffic from bridge building. Additionally, small, commonly used structures such as concrete pylons are in most cases prefabricated.

Radio towers for mobile phone and other services often consist of multiple prefabricated sections. Modern lattice towers and guyed masts are also commonly assembled of prefabricated elements.

Prefabrication has become widely used in the assembly of aircraft and spacecraft, with components such as wings and fuselage sections often being manufactured in different countries or states from the final assembly site. However this is sometimes for political rather than commercial reasons, such as for Airbus.

Advantages

1. Moving partial assemblies from a factory often costs less than moving pre-production resources to each site
2. Deploying resources on-site can add costs; prefabricating assemblies can save costs by reducing on-site work
3. Factory tools - jigs, cranes, conveyors, etc - can make production faster and more precise
4. Factory tools - shake tables, hydraulic testers, etc - can offer added quality assurance
5. Consistent indoor environments of factories eliminate most impacts of weather on production
6. Cranes and reusable factory supports can allow shapes and sequences without expensive on-site falsework
7. Higher-precision factory tools can aid more controlled movement of building heat & air, for lower energy consumption and healthier buildings
8. Factory production can facilitate more optimal materials usage, recycling, noise capture, dust capture, etc.
9. Machine-mediated parts movement, and freedom from wind & rain can improve construction safety

Disadvantages

1. Transportation costs may be higher for voluminous prefabricated sections than for their constituent materials, which can often be packed more densely.
2. Large prefabricated sections may require heavy-duty cranes and precision measurement and handling to place in position.

Off-site Fabrication

Off-Site fabrication is a process that incorporates prefabrication and pre-assembly. The process involves the design and manufacture of units or modules, usually remote from the work site, and the installation at the site to form the permanent works at the site. In its fullest sense, off-site fabrication requires a project strategy that will change the orientation of the project process from construction to manufacture to installation. Examples of off-site fabrication are wall panels for homes, wooden truss bridge spans, airport control stations.

There are 4 main categories of Off-Site fabrication, which is often also referred to as Off-Site Construction. These can be described as component (or sub-assembly) systems, penalised systems, volumetric systems, and modular systems. Below these categories different branches, or technologies are being developed. There are a vast number of different systems on the market which fall into these categories and with recent advances in digital design such as Building Information Modelling (BIM), the task of integrating these different systems into a construction project is becoming increasingly a "digital" management proposition.

References

- Malakooti, Behnam (2013). Operations and Production Systems with Multiple Objectives. John Wiley & Sons. ISBN 978-1-118-58537-5.
- Shewhart, Walter Andrew, Economic control of quality of manufactured product, 1931, New York: D. Van Nostrand Company. pp. 501 p.. ISBN 0-87389-076-0.
- Ptak, Carol & Smith, Chad (2011). Orlicky's MRP 3rd edition, McGraw Hill, New York ISBN 978-0-07-175563-4.
- Malakooti, Behnam (2013). Operations and Production Systems with Multiple Objectives. John Wiley & Sons. ISBN 978-1-118-58537-5.
- Zipkin Paul H., Foundations of Inventory Management, Boston: McGraw Hill, 2000, ISBN 0-256-11379-3
- Waldner, Jean-Baptiste (1992). "CIM: Principles of Computer Integrated Manufacturing". Chichester: John Wiley & Sons: 46. ISBN 0-471-93450-X.
- Smith, Debra & Smith, Chad (2013). Demand Driven Performance, Using Smart Metrics, McGraw Hill, ISBN 978-0071796095
- Mikell P. Groover (2007). Work Systems: The Methods, Measurement & Management of Work.

Prentice Hall. ISBN 978-0-13-140650-6

- Shigeo Shingo(1985). "A revolution in Manufacturing: The SMED System". Productivity Press. ISBN 0-915299-03-8
- "Moving Assembly Line at Ford". This Day in History. The History Channel. Retrieved September 2, 2016.
- Ament, Phil. "Assembly Line History: Invention of the Assembly Line". Ideafinder.com. Retrieved 2011-10-15.

3

Subdisciplines of Manufacturing Engineering

Mechanics, kinematics and technical drawing are some of the subdisciplines of manufacturing engineering. The text discusses the methods of manufacturing engineering and its subdisciplines in a critical manner providing key analysis to the subject matter.

Mechanics

Mechanics is an area of science concerned with the behaviour of physical bodies when subjected to forces or displacements, and the subsequent effects of the bodies on their environment. The scientific discipline has its origins in Ancient Greece with the writings of Aristotle and Archimedes. During the early modern period, scientists such as Khayaam, Galileo, Kepler, and Newton, laid the foundation for what is now known as classical mechanics. It is a branch of classical physics that deals with particles that are either at rest or are moving with velocities significantly less than the speed of light. It can also be defined as a branch of science which deals with the motion of and forces on objects.

Classical Versus Quantum Mechanics

Historically, classical mechanics came first, while quantum mechanics is a comparatively recent invention. Classical mechanics originated with Isaac Newton's laws of motion in *Principia Mathematica*; Quantum Mechanics was discovered in the early 20th century. Both are commonly held to constitute the most certain knowledge that exists about physical nature. Classical mechanics has especially often been viewed as a model for other so-called exact sciences. Essential in this respect is the relentless use of mathematics in theories, as well as the decisive role played by experiment in generating and testing them.

Quantum mechanics is of a wider scope, as it encompasses classical mechanics as a sub-discipline which applies under certain restricted circumstances. According to the correspondence principle, there is no contradiction or conflict between the two subjects, each simply pertains to specific situations. The correspondence principle states that the behavior of systems described by quantum theories reproduces classical physics in the limit of large quantum numbers. Quantum mechanics has superseded classical mechanics at the foundational level and is indispensable for the explanation and prediction of processes at the molecular, atomic, and sub-atomic level. However, for macroscopic processes classical mechanics is able to solve problems which are unmanageably difficult in quantum mechanics and hence remains useful and well used. Modern descriptions of such behavior begin with a careful definition of such quantities as displacement (distance moved), time, velocity, acceleration, mass, and force. Until about 400 years ago, however, motion was explained from a very different point of view. For example, following the ideas of Greek philosopher and scientist Aristotle, scientists reasoned that a cannonball falls down because its natural position is in the Earth; the sun, the moon, and the stars travel in circles around the earth because it is the nature of heavenly objects to travel in perfect circles.

Often cited as the father of modern science, Galileo brought together the ideas of other great thinkers of his time and began to analyze motion in terms of distance traveled from some starting position and the time that it took. He showed that the speed of falling objects increases steadily during the time of their fall. This acceleration is the same for heavy objects as for light ones, provided air friction (air resistance) is discounted. The English mathematician and physicist Isaac Newton improved this analysis by defining force and mass and relating these to acceleration. For objects traveling at speeds close to the speed of light, Newton's laws were superseded by Albert Einstein's theory of relativity. For atomic and subatomic particles, Newton's laws were superseded by quantum theory. For everyday phenomena, however, Newton's three laws of motion remain the cornerstone of dynamics, which is the study of what causes motion.

Relativistic Versus Newtonian Mechanics

In analogy to the distinction between quantum and classical mechanics, Einstein's general and special theories of relativity have expanded the scope of Newton and Galileo's formulation of mechanics. The differences between relativistic and Newtonian mechanics become significant and even dominant as the velocity of a massive body approaches the speed of light. For instance, in Newtonian mechanics, Newton's laws of motion specify that $\mathbf{F} = m\mathbf{a}$, whereas in Relativistic mechanics and Lorentz transformations, which were first discovered by Hendrik Lorentz, $\mathbf{F} = \gamma m\mathbf{a}$ (where γ is the Lorentz factor, which is almost equal to 1 for low speeds).

General Relativistic Versus Quantum

Relativistic corrections are also needed for quantum mechanics, although general relativity has not been integrated. The two theories remain incompatible, a hurdle which must be overcome in developing a theory of everything.

History

Antiquity

The main theory of mechanics in antiquity was Aristotelian mechanics. A later developer in this tradition is Hipparchus.

Medieval Age

Arabic Machine Manuscript. Unknown date (at a guess: 16th to 19th centuries).

In the Middle Ages, Aristotle's theories were criticized and modified by a number of figures, beginning with John Philoponus in the 6th century. A central problem was that of projectile motion, which was discussed by Hipparchus and Philoponus. This led to the development of the theory of impetus by 14th-century French priest Jean Buridan, which developed into the modern theories of inertia, velocity, acceleration and momentum. This work and others was developed in 14th-century England by the Oxford Calculators such as Thomas Bradwardine, who studied and formulated various laws regarding falling bodies.

On the question of a body subject to a constant (uniform) force, the 12th-century Jewish-Arab Nathanel (Iraqi, of Baghdad) stated that constant force imparts constant acceleration, while the main properties are uniformly accelerated motion (as of falling bodies) was worked out by the 14th-century Oxford Calculators.

Early Modern Age

Two central figures in the early modern age are Galileo Galilei and Isaac Newton. Galileo's final statement of his mechanics, particularly of falling bodies, is his *Two New Sciences* (1638). Newton's 1687 *Philosophiæ Naturalis Principia Mathematica* provided a detailed mathematical account of mechanics, using the newly developed mathematics of calculus and providing the basis of Newtonian mechanics.

There is some dispute over priority of various ideas: Newton's *Principia* is certainly the seminal work and has been tremendously influential, and the systematic mathematics therein did not and could not have been stated earlier because calculus had not been developed. However, many of the ideas, particularly as pertain to inertia (impetus) and falling bodies had been developed and stated by earlier researchers, both the then-recent Galileo and the less-known medieval predecessors. Precise credit is at times difficult or contentious because scientific language and standards of proof changed, so whether medieval statements are *equivalent* to modern statements or *sufficient* proof, or instead *similar* to modern statements and *hypotheses* is often debatable.

Modern Age

Two main modern developments in mechanics are general relativity of Einstein, and quantum mechanics, both developed in the 20th century based in part on earlier 19th-century ideas. The development in the modern continuum mechanics, particularly in the areas of elasticity, plasticity, fluid dynamics, electrodynamics and thermodynamics of deformable media, started in the second half of the 20th century.

Types of Mechanical Bodies

The often-used term body needs to stand for a wide assortment of objects, including particles, projectiles, spacecraft, stars, parts of machinery, parts of solids, parts of fluids (gases and liquids), etc.

Other distinctions between the various sub-disciplines of mechanics, concern the nature of the bodies being described. Particles are bodies with little (known) internal structure, treated as mathematical points in classical mechanics. Rigid bodies have size and shape, but retain a simplicity close to that of the particle, adding just a few so-called degrees of freedom, such as orientation in space.

Otherwise, bodies may be semi-rigid, i.e. elastic, or non-rigid, i.e. fluid. These subjects have both classical and quantum divisions of study.

For instance, the motion of a spacecraft, regarding its orbit and attitude (rotation), is described by the relativistic theory of classical mechanics, while the analogous movements of an atomic nucleus are described by quantum mechanics.

Sub-disciplines in Mechanics

The following are two lists of various subjects that are studied in mechanics.

Note that there is also the "theory of fields" which constitutes a separate discipline in physics, formally treated as distinct from mechanics, whether classical fields or quantum fields. But in actual practice, subjects belonging to mechanics and fields are closely interwoven. Thus, for instance, forces that act on particles are frequently derived from fields (electromagnetic or gravitational), and particles generate fields by acting as sources. In fact, in quantum mechanics, particles themselves are fields, as described theoretically by the wave function.

Classical Mechanics

The following are described as forming classical mechanics:

- Newtonian mechanics, the original theory of motion (kinematics) and forces (dynamics).
- Analytical mechanics is a reformulation of Newtonian mechanics with an emphasis on system energy, rather than on forces. There are two main branches of analytical mechanics:
    - Hamiltonian mechanics, a theoretical formalism, based on the principle of conservation of energy.
    - Lagrangian mechanics, another theoretical formalism, based on the principle of the least action.
- Classical statistical mechanics generalizes ordinary classical mechanics to consider systems in an unknown state; often used to derive thermodynamic properties.
- Celestial mechanics, the motion of bodies in space: planets, comets, stars, galaxies, etc.
- Astrodynamics, spacecraft navigation, etc.
- Solid mechanics, elasticity, the properties of deformable bodies.
- Fracture mechanics
- Acoustics, sound (= density variation propagation) in solids, fluids and gases.
- Statics, semi-rigid bodies in mechanical equilibrium
- Fluid mechanics, the motion of fluids
- Soil mechanics, mechanical behavior of soils
- Continuum mechanics, mechanics of continua (both solid and fluid)
- Hydraulics, mechanical properties of liquids
- Fluid statics, liquids in equilibrium
- Applied mechanics, or Engineering mechanics

- Biomechanics, solids, fluids, etc. in biology
- Biophysics, physical processes in living organisms
- Relativistic or Einsteinian mechanics, universal gravitation.

Quantum Mechanics

The following are categorized as being part of quantum mechanics:

- Schrödinger wave mechanics, used to describe the motion of the wavefunction of a single particle.
- Matrix mechanics is an alternative formulation that allows considering systems with a finite-dimensional state space.
- Quantum statistical mechanics generalizes ordinary quantum mechanics to consider systems in an unknown state; often used to derive thermodynamic properties.
- Particle physics, the motion, structure, and reactions of particles
- Nuclear physics, the motion, structure, and reactions of nuclei
- Condensed matter physics, quantum gases, solids, liquids, etc.

Professional Organizations

- Applied Mechanics Division, American Society of Mechanical Engineers
- Fluid Dynamics Division, American Physical Society
- Society for Experimental Mechanics
- Institution of Mechanical Engineers is the United Kingdom's qualifying body for Mechanical Engineers and has been the home of Mechanical Engineers for over 150 years.
- International Union of Theoretical and Applied Mechanics

Kinematics

Kinematics is the branch of classical mechanics which describes the motion of points (alternatively "particles"), bodies (objects), and systems of bodies without consideration of the masses of those objects nor the forces that may have caused the motion. Kinematics as a field of study is often referred to as the "geometry of motion" and as such may be seen as a branch of mathematics. Kinematics begins with a description of the geometry of the system and the initial conditions of known values of the position, velocity and or acceleration of various points that are a part of the system, then from geometrical arguments it can determine the position, the velocity and the acceleration of any part of the system. The study of the influence of forces acting on masses falls within the purview of kinetics.

Kinematics is used in astrophysics to describe the motion of celestial bodies and collections of such bodies. In mechanical engineering, robotics, and biomechanics kinematics is used to describe the motion of systems composed of joined parts (multi-link systems) such as an engine, a robotic arm or the skeleton of the human body.

The use of geometric transformations, also called rigid transformations, to describe the movement of components of a mechanical system simplifies the derivation of its equations of motion, and is central to dynamic analysis.

Kinematic analysis is the process of measuring the kinematic quantities used to describe motion. In engineering, for instance, kinematic analysis may be used to find the range of movement for a given mechanism, and working in reverse, using kinematic synthesis used to design a mechanism for a desired range of motion. In addition, kinematics applies algebraic geometry to the study of the mechanical advantage of a mechanical system or mechanism.

Kinematic and cinématique are related to the French word cinéma, but neither are directly derived from it. However, they do share a root word in common, as cinéma came from the shortened form of cinématographe, "motion picture projector and camera," once again from the Greek word for movement but also the Greek word for writing.

Kinematics of a Particle Trajectory in a Non-rotating Frame of Reference

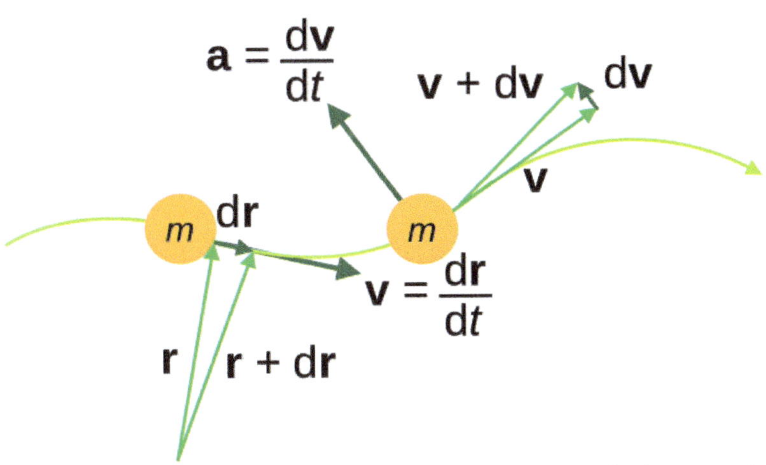

Kinematic quantities of a classical particle: mass m, position $\mathbf{r}$, velocity $\mathbf{v}$, acceleration $\mathbf{a}$.

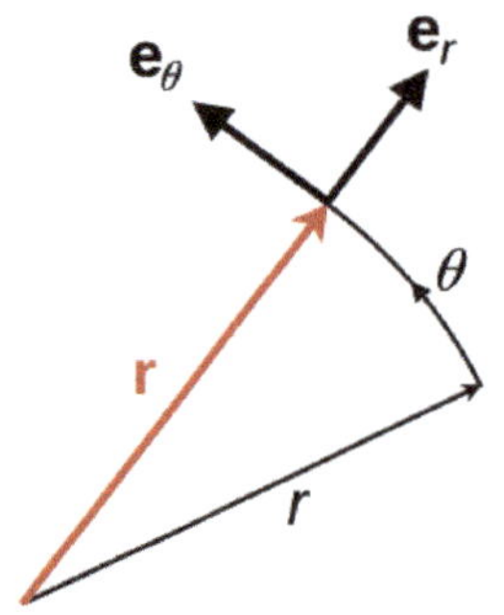

Position vector **r**, always points radially from the origin.

Kinematic vectors in plane polar coordinates. Notice the setup is not restricted to 2d space, but a plane in any higher dimension.

Particle kinematics is the study of the trajectory of a particle. The position of a particle is defined to be the coordinate vector from the origin of a coordinate frame to the particle. For example, consider a tower 50 m south from your home, where the coordinate frame is located at your home, such that East is the x-direction and North is the y-direction, then the coordinate vector to the base of the tower is **r**=(0, -50, 0). If the tower is 50 m high, then the coordinate vector to the top of the tower is **r**=(0, -50, 50).

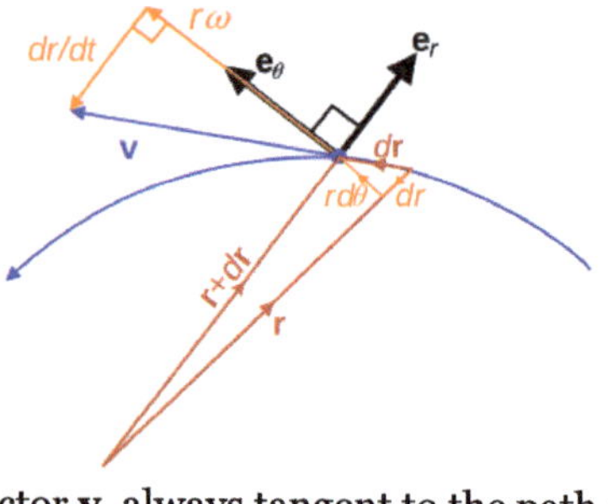

Velocity vector **v**, always tangent to the path of motion.

In the most general case, a three-dimensional coordinate system is used to define the position of a particle. However, if the particle is constrained to move in a surface, a two-dimensional coordinate system is sufficient. All observations in physics are incomplete without those observations being described with respect to a reference frame.

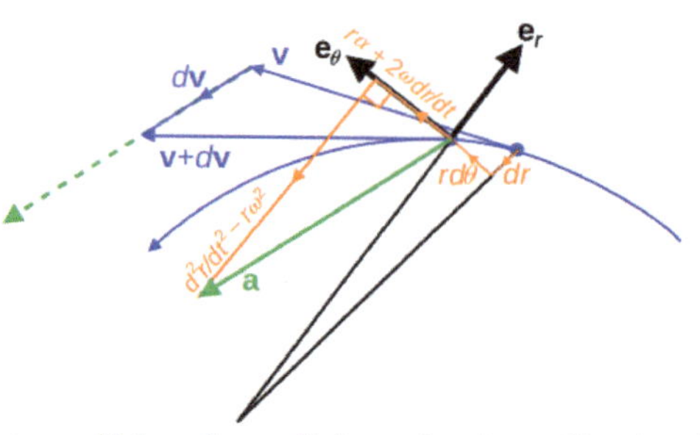

Acceleration vector a, not parallel to the radial motion but offset by the angular and Coriolis accelerations, nor tangent to the path but offset by the centripetal and radial accelerations.

The position vector of a particle is a vector drawn from the origin of the reference frame to the particle. It expresses both the distance of the point from the origin and its direction from the origin. In three dimensions, the position of point P can be expressed as

$$\mathbf{P} = (x_P, y_P, z_P) = x_P\hat{i} + y_P\hat{j} + z_P\hat{k},$$

where x_p, y_p, and z_p are the Cartesian coordinates and i, j and k are the unit vectors along the x, y, and z coordinate axes, respectively. The magnitude of the position vector $|\mathbf{P}|$ gives the distance between the point P and the origin.

$$|\mathbf{P}| = \sqrt{x_P^2 + y_P^2 + z_P^2}.$$

The direction cosines of the position vector provide a quantitative measure of direction. It is important to note that the position vector of a particle isn't unique. The position vector of a given particle is different relative to different frames of reference.

The *trajectory* of a particle is a vector function of time, $\mathbf{P}(t)$, which defines the curve traced by the moving particle, given by

$$\mathbf{P}(t) = x_P(t)\hat{i} + y_P(t)\hat{j} + z_P(t)\hat{k},$$

where the coordinates x_p, y_p, and z_p are each functions of time.

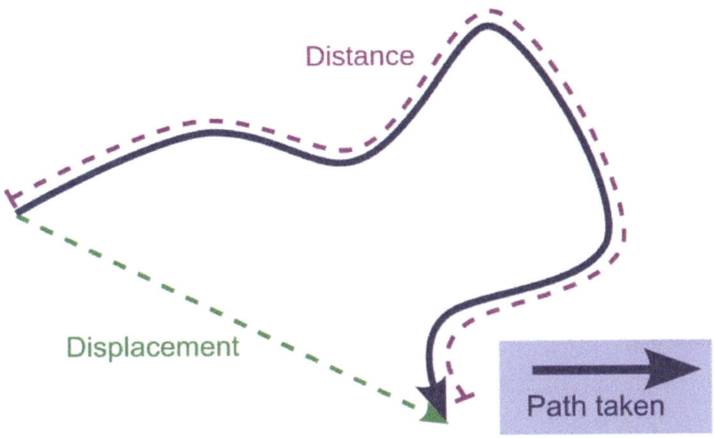

The distance travelled is always greater than or equal to the displacement.

Velocity and Speed

The velocity of a particle is a vector quantity that describes the direction of motion and the magnitude of the motion of particle. More mathematically, the rate of change of the position vector of a point, with respect to time is the velocity of the point. Consider the ratio of the difference of two positions of a particle divided by the time interval, which is called the average velocity over that time interval. This average velocity is defined as

$$\overline{\mathbf{V}} = \frac{\Delta \mathbf{P}}{\Delta t},$$

where $\Delta\mathbf{P}$ is the change in the position vector over the time interval Δt.

In the limit as the time interval Δt becomes smaller and smaller, the average velocity becomes the time derivative of the position vector,

$$\mathbf{V} = \lim_{\Delta t \to 0} \frac{\Delta \mathbf{P}}{\Delta t} = \frac{d\mathbf{P}}{dt} = \dot{\mathbf{P}} = \dot{x}_p \hat{i} + \dot{y}_p \hat{j} + \dot{z}_p \hat{k}.$$

Thus, velocity is the time rate of change of position of a point, and the dot denotes the derivative of those functions x, y, and z with respect to time. Furthermore, the velocity is tangent to the trajectory of the particle at every position the particle occupies along its path. Note that in a non-rotating frame of reference, the derivatives of the coordinate directions are not considered as their directions and magnitudes are constants.

The speed of an object is the magnitude $|\mathbf{V}|$ of its velocity. It is a scalar quantity:

$$|\mathbf{V}| = |\dot{\mathbf{P}}| = \frac{ds}{dt},$$

where s is the arc-length measured along the trajectory of the particle. This arc-length traveled by a particle over time is a non-decreasing quantity. Hence, ds/dt is non-negative, which implies that speed is also non-negative.

Acceleration

The velocity vector can change in magnitude and in direction or both at once. Hence, the acceleration is the rate of change of the magnitude of the velocity vector plus the rate of change of direction of that vector. The same reasoning used with respect to the position of a particle to define velocity, can be applied to the velocity to define acceleration. The acceleration of a particle is the vector defined by the rate of change of the velocity vector. The average acceleration of a particle over a time interval is defined as the ratio.

$$\bar{\mathbf{A}} = \frac{\Delta \mathbf{V}}{\Delta t},$$

where $\Delta\mathbf{V}$ is the difference in the velocity vector and Δt is the time interval.

The acceleration of the particle is the limit of the average acceleration as the time interval approaches zero, which is the time derivative,

$$\text{Eqn 1)}\quad \mathbf{A} = \lim_{\Delta t \to 0} \frac{\Delta \mathbf{V}}{\Delta t} = \frac{d\mathbf{V}}{dt} = \dot{\mathbf{V}} = \dot{v}_x \hat{i} + \dot{v}_y \hat{j} + \dot{v}_z \hat{k}$$

or

$$\mathbf{A} = \ddot{\mathbf{P}} = \ddot{x}_p \hat{i} + \ddot{y}_p \hat{j} + \ddot{z}_p \hat{k}$$

Thus, acceleration is the first derivative of the velocity vector and the second derivative of the position vector of that particle. Note that in a non-rotating frame of reference, the derivatives of the coordinate directions are not considered as their directions and magnitudes are constants.

The magnitude of the acceleration of an object is the magnitude |**A**| of its acceleration vector. It is a scalar quantity:

$$|\mathbf{A}| = |\dot{\mathbf{V}}| = \frac{dv}{dt},$$

Relative Position Vector

A relative position vector is a vector that defines the position of one point relative to another. It is the difference in position of the two points. The position of one point *A* relative to another point *B* is simply the difference between their positions

$$\mathbf{P}_{A/B} = \mathbf{P}_A - \mathbf{P}_B$$

which is the difference between the components of their position vectors.

If point *A* has position components $\mathbf{P}_A = (X_A, Y_A, Z_A)$

If point *B* has position components $\mathbf{P}_B = (X_B, Y_B, Z_B)$

then the position of point *A* relative to point *B* is the difference between their components: $\mathbf{P}_{A/B} = \mathbf{P}_A - \mathbf{P}_B = (X_A - X_B, Y_A - Y_B, Z_A - Z_B)$

Relative Velocity

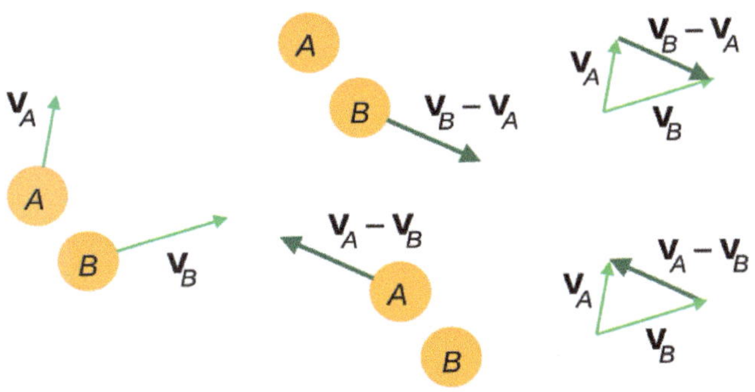

Relative velocities between two particles in classical mechanics.

The velocity of one point relative to another is simply the difference between their velocities

$$\mathbf{V}_{A/B} = \mathbf{V}_A - \mathbf{V}_B$$

which is the difference between the components of their velocities.

If point *A* has velocity components $\mathbf{V}_A = (V_{A_x}, V_{A_y}, V_{A_z})$

and point *B* has velocity components $\mathbf{V}_B = (V_{B_x}, V_{B_y}, V_{B_z})$

then the velocity of point *A* relative to point *B* is the difference between their components: $\mathbf{V}_{A/B} = \mathbf{V}_A - \mathbf{V}_B = (V_{A_x} - V_{B_x}, V_{A_y} - V_{B_y}, V_{A_z} - V_{B_z})$

Alternatively, this same result could be obtained by computing the time derivative of the relative position vector $\mathbf{R}_{B/A}$.

In the case where the velocity is close to the speed of light *c* (generally within 95%), another scheme of relative velocity called rapidity, that depends on the ratio of V to c, is used in special relativity.

Relative Acceleration

The acceleration of one point *C* relative to another point *B* is simply the difference between their accelerations.

$$\mathbf{A}_{C/B} = \mathbf{A}_C - \mathbf{A}_B$$

which is the difference between the components of their accelerations.

If point *C* has acceleration components $\mathbf{A}_C = \left(A_{C_x}, A_{C_y}, A_{C_z}\right)$

and point *B* has acceleration components $\mathbf{A}_B = \left(A_{B_x}, A_{B_y}, A_{B_z}\right)$

then the acceleration of point *C* relative to point *B* is the difference between their components: $\mathbf{A}_{C/B} = \mathbf{A}_C - \mathbf{A}_B = \left(A_{C_x} - A_{B_x}, A_{C_y} - A_{B_y}, A_{C_z} - A_{B_z}\right)$

Alternatively, this same result could be obtained by computing the second time derivative of the relative position vector $\mathbf{P}_{B/A}$.

Particle Trajectories Under Constant Acceleration

For the case of constant acceleration, the differential equation Eq 1) can be integrated as the acceleration vector A of a point P is constant in magnitude and direction. Such a point is said to be undergoing uniformly accelerated motion. In this case, the velocity V(t) and then the trajectory P(t) of the particle can be obtained by integrating the acceleration equation A with respect to time.

Assuming that the initial conditions of the position, $\mathbf{P}_0$, and velocity $\mathbf{V}_0$ are known, the first integration yields the velocity of the particle as a function of time.

$$\mathbf{V}(t) = \int_0^t \mathbf{A}dt = \mathbf{A}t + \mathbf{V}_0.$$

A second integration yields its path (trajectory),

$$\mathbf{P}(t) = \int_0^t \mathbf{V}(t)dt = \int (\mathbf{A}t + \mathbf{V}_0)dt = \tfrac{1}{2}\mathbf{A}t^2 + \mathbf{V}_0 t + \mathbf{P}_0.$$

Additional relations between displacement, velocity, acceleration, and time can be derived. Since the acceleration is constant,

$\mathbf{A} = \dfrac{\Delta \mathbf{V}}{\Delta t} = \dfrac{\mathbf{V} - \mathbf{V}_0}{t}$ can be substituted into the above equation to give:

$$\mathbf{P}(t) = \mathbf{P}_0 + \left(\frac{\mathbf{V}+\mathbf{V}_0}{2}\right)t.$$

A relationship between velocity, position and acceleration without explicit time dependence can be had by solving the average acceleration for time and substituting and simplifying

$$t = \frac{\mathbf{V}-\mathbf{V}_0}{\mathbf{A}}$$

$$(\mathbf{P}-\mathbf{P}_0)^\circ\mathbf{A} = (\mathbf{V}-\mathbf{V}_0)^\circ\frac{\mathbf{V}+\mathbf{V}_0}{2},$$

where c denotes the dot product, which is appropriate as the products are scalars rather than vectors.

$$2(\mathbf{P}-\mathbf{P}_0)^\circ\mathbf{A} = |\mathbf{V}|^2 - |\mathbf{V}_0|^2.$$

The dot can be replaced by the cosine of the angle α between the vectors and the vectors by their magnitudes, in which case:

$$2(|\mathbf{P}|-|\mathbf{P}_0|)\cdot|\mathbf{A}|\cdot \cos\alpha = |\mathbf{V}|^2 - |\mathbf{V}_0|^2.$$

In the case of acceleration always in the direction of the motion, the angle between the vectors (α) = 0, cos(0) = 1 and,

$$|\mathbf{V}|^2 = |\mathbf{V}_0|^2 + 2|\mathbf{A}|(|\mathbf{P}-\mathbf{P}_0|).$$

This can be simplified using the notation for the magnitudes of the vectors $|\mathbf{A}| = a, |\mathbf{V}| = v, |\mathbf{P}-\mathbf{P}_0| = \delta s$ where δs can be any curvaceous path taken as the constant tangential acceleration is applied along that path, so

$$v^2 = v_0^2 + 2a(\delta s).$$

This reduces the parametric equations of motion of the particle to a cartesian relationship of speed versus position. This relation is useful when time is not known explicitly.

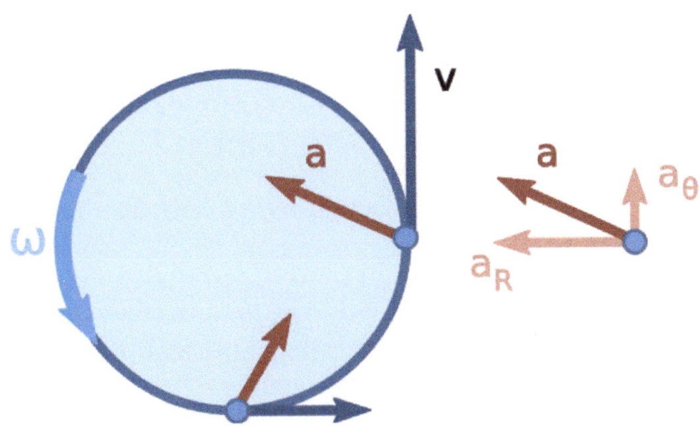

Velocity and acceleration for nonuniform circular motion: the velocity vector is tangential to the orbit, but the acceleration vector is not radially inward because of its tangential component $\mathbf{a}_\theta$ that increases the rate of rotation: $d\omega/dt = |\mathbf{a}_\theta|/R$.

Particle Trajectories in Cylindrical-polar Coordinates

It is often convenient to formulate the trajectory of a particle **P**(t) = (X(t), Y(t) and Z(t)) using polar coordinates in the *X-Y* plane. In this case, its velocity and acceleration take a convenient form.

Recall that the trajectory of a particle P is defined by its coordinate vector P measured in a fixed reference frame F. As the particle moves, its coordinate vector P(t) traces its trajectory, which is a curve in space, given by:

$$\mathbf{P}(t) = X(t)\hat{\imath} + Y(t)\hat{\jmath} + Z(t)\hat{k},$$

where *i*, *j*, and *k* are the unit vectors along the *X*, *Y* and *Z* axes of the reference frame *F*, respectively.

Consider a particle *P* that moves only on the surface of a circular cylinder R(t)=constant, it is possible to align the *Z* axis of the fixed frame *F* with the axis of the cylinder. Then, the angle θ around this axis in the *X-Y* plane can be used to define the trajectory as,

$$\mathbf{P}(t) = R\cos\theta(t)\hat{\imath} + R\sin\theta(t)\hat{\jmath} + Z(t)\hat{k}.$$

The cylindrical coordinates for P(t) can be simplified by introducing the radial and tangential unit vectors,

$$\mathbf{e}_r = \cos\theta(t)\hat{\imath} + \sin\theta(t)\hat{\jmath}, \quad \mathbf{e}_t = -\sin\theta(t)\hat{\imath} + \cos\theta(t)\hat{\jmath}.$$

and their time derivatives from elementary calculus:

$$\frac{d}{dt}\mathbf{e}_r = \dot{\mathbf{e}}_r = \dot{\theta}\mathbf{e}_t$$

$$\frac{d}{dt}\dot{\mathbf{e}}_r = \ddot{\mathbf{e}}_r = \ddot{\theta}\mathbf{e}_t - \dot{\theta}\mathbf{e}_r$$

$$\frac{d}{dt}\mathbf{e}_t = \dot{\mathbf{e}}_t = -\dot{\theta}\mathbf{e}_r\,.$$

$$\frac{d}{dt}\dot{\mathbf{e}}_t = \ddot{\mathbf{e}}_t = -\ddot{\theta}\mathbf{e}_r - \dot{\theta}^2\mathbf{e}_t.\ \frac{d}{dt}\dot{\mathbf{e}}_t = \ddot{\mathbf{e}}_t = -\ddot{\theta}\mathbf{e}_r - \dot{\theta}^2\mathbf{e}_t.$$

Using this notation, P(t) takes the form,

$$\mathbf{P}(t) = R\mathbf{e}_r + Z(t)\hat{k},$$

where *R* is constant in the case of the particle moving only on the surface of a cylinder of radius *R*.

In general, the trajectory P(t) is not constrained to lie on a circular cylinder, so the radius *R* varies with time and the trajectory of the particle in cylindrical-polar coordinates becomes:

$$\mathbf{P}(t) = R(t)\mathbf{e}_r + Z(t)\hat{k}.$$

Where R, theta, and Z might be continuously differentiable functions of time and the function notation is dropped for simplicity. The velocity vector $\mathbf{V}_P$ is the time derivative of the trajectory P(t), which yields:

$$\mathbf{V}_P = \frac{d}{dt}(R\mathbf{e}_r + Z\hat{k}) = \dot{R}\mathbf{e}_r + R\dot{\mathbf{e}}_r + \dot{Z}\hat{k} = \dot{R}\mathbf{e}_r + R\dot{\theta}\mathbf{e}_t + \dot{Z}\hat{k}.$$

Similarly, the acceleration $\mathbf{A}_P$, which is the time derivative of the velocity $\mathbf{V}_P$, is given by:

$$\mathbf{A}_P = \frac{d}{dt}(\dot{R}\mathbf{e}_r + R\dot{\theta}\mathbf{e}_t + \dot{Z}\hat{k}) = (\ddot{R} - R\dot{\theta}^2)\mathbf{e}_r + (R\ddot{\theta} + 2\dot{R}\dot{\theta})\mathbf{e}_t + \ddot{Z}\hat{k}.$$

The term $--R\dot{\theta}^2\mathbf{e}_r$ acts toward the center of curvature of the path at that point on the path, is commonly called the centripetal acceleration. The term $2\dot{R}\dot{\theta}\mathbf{e}_t$ is called the Coriolis acceleration.

Constant Radius

If the trajectory of the particle is constrained to lie on a cylinder, then the radius R is constant and the velocity and acceleration vectors simplify. The velocity of $\mathbf{V}_P$ is the time derivative of the trajectory **P**(t),

$$\mathbf{V}_P = \frac{d}{dt}(R\mathbf{e}_r + Z\hat{k}) = R\dot{\theta}\mathbf{e}_t + \dot{Z}\hat{k}.$$

The acceleration vector becomes:

$$\mathbf{A}_P = \frac{d}{dt}(R\dot{\theta}\mathbf{e}_t + \dot{Z}\hat{k}) = -R\dot{\theta}^2\mathbf{e}_r + R\ddot{\theta}\mathbf{e}_t + \ddot{Z}\hat{k}.$$

Planar Circular Trajectories

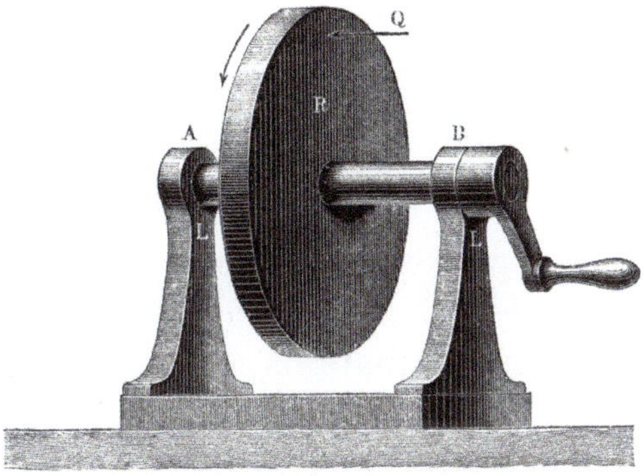

Each particle on the wheel travels in a planar circular trajectory (Kinematics of Machinery, 1876).

A special case of a particle trajectory on a circular cylinder occurs when there is no movement along the Z axis:

$$\mathbf{P}(t) = R\mathbf{e}_r + Z_0\hat{k},$$

where R and Z_0 are constants. In this case, the velocity $\mathbf{V}_P$ is given by:

$$\mathbf{V}_P = \frac{d}{dt}(R\mathbf{e}_r + Z_0\hat{k}) = R\dot{\theta}\mathbf{e}_t = R\omega\mathbf{e}_t,$$

where

$$\omega = \dot{\theta},$$

is the angular velocity of the unit vector $\mathbf{e}_t$ around the z axis of the cylinder.

The acceleration $\mathbf{A}_P$ of the particle P is now given by:

$$\mathbf{A}_P = \frac{d}{dt}(R\dot{\theta}\mathbf{e}_t) = -R\dot{\theta}^2\mathbf{e}_r + R\ddot{\theta}\mathbf{e}_t.$$

The components

$$a_r = -R\dot{\theta}^2, \quad a_t = R\ddot{\theta},$$

are called, respectively, the *radial* and *tangential components* of acceleration.

The notation for angular velocity and angular acceleration is often defined as

$$\omega = \dot{\theta}, \quad \alpha = \ddot{\theta},$$

so the radial and tangential acceleration components for circular trajectories are also written as

$$a_r = -R\omega^2, \quad a_t = R\alpha.$$

Point Trajectories in a Body Moving in the Plane

The movement of components of a mechanical system are analyzed by attaching a reference frame to each part and determining how the various reference frames move relative to each other. If the structural stiffness of the parts are sufficient, then their deformation can be neglected and rigid transformations can be used to define this relative movement. This reduces the description of the motion of the various parts of a complicated mechanical system to a problem of describing the geometry of each part and geometric association of each part relative to other parts.

Geometry is the study of the properties of figures that remain the same while the space is transformed in various ways---more technically, it is the study of invariants under a set of transformations. These transformations can cause the displacement of the triangle in the plane, while leaving the vertex angle and the distances between vertices unchanged. Kinematics is often described as applied geometry, where the movement of a mechanical system is described using the rigid transformations of Euclidean geometry.

The coordinates of points in a plane are two-dimensional vectors in $\mathbf{R}^2$ (two dimensional space). Rigid transformations are those that preserve the distance between any two points. The set of rigid transformations in an n-dimensional space is called the special Euclidean group on R^n, and denoted $SE(n)$.

Displacements and Motion

The position of one component of a mechanical system relative to another is defined by introducing a reference frame, say *M*, on one that moves relative to a fixed frame, *F*, on the other. The rigid transformation, or displacement, of *M* relative to *F* defines the relative position of the two components. A displacement consists of the combination of a rotation and a translation.

The set of all displacements of *M* relative to *F* is called the configuration space of *M*. A smooth curve from one position to another in this configuration space is a continuous set of displacements, called the motion of *M* relative to *F*. The motion of a body consists of a continuous set of rotations and translations.

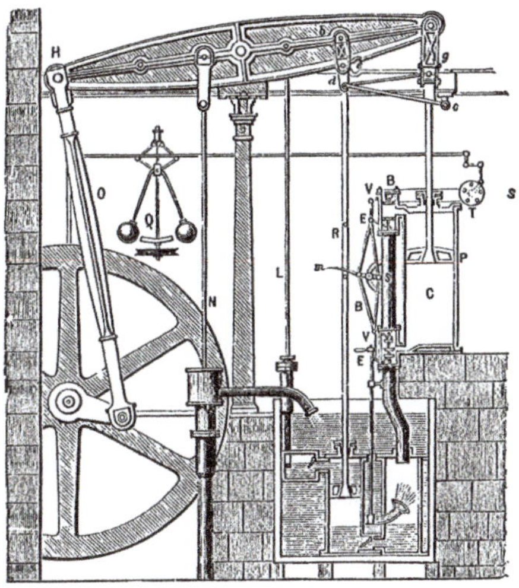

The movement of each of the components of the Boulton & Watt Steam Engine (1784) is modeled by a continuous set of rigid displacements.

Matrix Representation

The combination of a rotation and translation in the plane $\mathbf{R}^2$ can be represented by a certain type of 3x3 matrix known as a homogeneous transform. The 3x3 homogeneous transform is constructed from a 2x2 rotation matrix A(φ) and the 2x1 translation vector $\mathbf{d}=(d_x, d_y)$, as:

$$[T(\phi,\mathbf{d})] = \begin{bmatrix} A(\phi) & \mathbf{d} \\ 0 & 1 \end{bmatrix} = \begin{bmatrix} \cos\phi & -\sin\phi & d_x \\ \sin\phi & \cos\phi & d_y \\ 0 & 0 & 1 \end{bmatrix}.$$

These homogeneous transforms perform rigid transformations on the points in the plane z=1, that is on points with coordinates $\mathbf{p}=(x, y, 1)$.

In particular, let p define the coordinates of points in a reference frame M coincident with a fixed frame F. Then, when the origin of M is displaced by the translation vector d relative to the origin of F and rotated by the angle φ relative to the x-axis of F, the new coordinates in F of points in M are given by:

$$\mathbf{P} = [T(\phi,\mathbf{d})]\mathbf{p} = \begin{bmatrix} \cos\phi & -\sin\phi & d_x \\ \sin\phi & \cos\phi & d_y \\ 0 & 0 & 1 \end{bmatrix} \begin{Bmatrix} x \\ y \\ 1 \end{Bmatrix}.$$

Homogeneous transforms represent affine transformations. This formulation is necessary because a translation is not a linear transformation of R^2. However, using projective geometry, so that R^2 is considered to be a subset of R^3, translations become affine linear transformations.

Pure Translation

If a rigid body moves so that its reference frame *M* does not rotate (Ø=0) relative to the fixed frame *F*, the motion is said to be pure translation. In this case, the trajectory of every point in the body is an offset of the trajectory **d**(t) of the origin of *M*, that is:

$$\mathbf{P}(t) = [T(0, \mathbf{d}(t))]\mathbf{p} = \mathbf{d}(t) + \mathbf{p}.$$

Thus, for bodies in pure translation, the velocity and acceleration of every point *P* in the body are given by:

$$\mathbf{V}_P = \dot{\mathbf{P}}(t) = \dot{\mathbf{d}}(t) = \mathbf{V}_O, \quad \mathbf{A}_P = \ddot{\mathbf{P}}(t) = \ddot{\mathbf{d}}(t) = \mathbf{A}_O,$$

where the dot denotes the derivative with respect to time and V_O and A_O are the velocity and acceleration, respectively, of the origin of the moving frame M. Recall the coordinate vector p in M is constant, so its derivative is zero.

Rotation of a Body Around a Fixed Axis

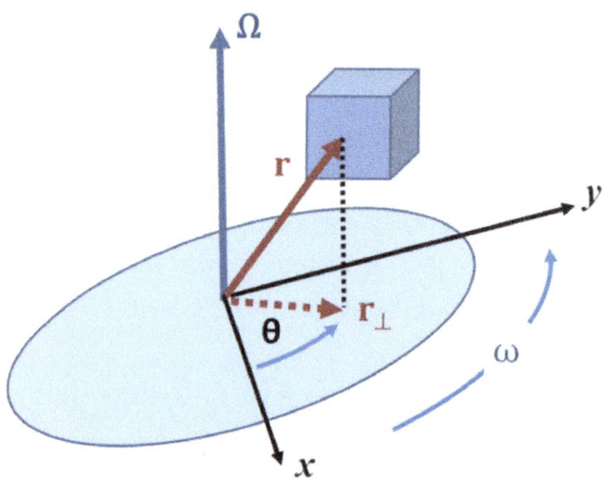

The angular velocity vector **Ω** points up for counterclockwise rotation and down for clockwise rotation, as specified by the right-hand rule. Angular position $\theta(t)$ changes with time at a rate $\omega(t) = d\theta/dt$.

Rotational or angular kinematics is the description of the rotation of an object. The description of rotation requires some method for describing orientation. Common descriptions include Euler angles and the kinematics of turns induced by algebraic products.

In what follows, attention is restricted to simple rotation about an axis of fixed orientation. The *z*-axis has been chosen for convenience.

Position

This allows the description of a rotation as the angular position of a planar reference frame M relative to a fixed F about this shared z-axis. Coordinates p=(x, y) in M are related to coordinates P=(X, Y) in F by the matrix equation:

$$\mathbf{P}(t) = [A(t)]\mathbf{p},$$

where

$$[A(t)] = \begin{bmatrix} \cos\theta(t) & -\sin\theta(t) \\ \sin\theta(t) & \cos\theta(t) \end{bmatrix},$$

is the rotation matrix that defines the angular position of M relative to F as a function of time.

Velocity

If the point p does not move in M, its velocity in F is given by

$$\mathbf{V}_P = \dot{\mathbf{P}} = [\dot{A}(t)]\mathbf{p}.$$

It is convenient to eliminate the coordinates **p** and write this as an operation on the trajectory P(t),

$$\mathbf{V}_P = [\dot{A}(t)][A(t)^{-1}]\mathbf{P} = [\Omega]\mathbf{P},$$

where the matrix

$$[\Omega] = \begin{bmatrix} 0 & -\omega \\ \omega & 0 \end{bmatrix},$$

is known as the angular velocity matrix of M relative to F. The parameter ω is the time derivative of the angle θ, that is:

$$\omega = \frac{d\theta}{dt}.$$

Acceleration

The acceleration of **P**(t) in F is obtained as the time derivative of the velocity,

$$\mathbf{A}_P = \ddot{P}(t) = [\dot{\Omega}]\mathbf{P} + [\Omega]\dot{\mathbf{P}},$$

which becomes

$$\mathbf{A}_P = [\dot{\Omega}]\mathbf{P} + [\Omega][\Omega]\mathbf{P},$$

where

$$[\dot{\Omega}] = \begin{bmatrix} 0 & -\alpha \\ \alpha & 0 \end{bmatrix},$$

is the angular acceleration matrix of M on F, and

$$\alpha = \frac{d^2\theta}{dt^2}.$$

The description of rotation then involves these three quantities:

- Angular position : the oriented distance from a selected origin on the rotational axis to a point of an object is a vector **r** (t) locating the point. The vector $\mathbf{r}(t)$ has some projection (or, equivalently, some component) $\mathbf{r}_\perp(t)$ on a plane perpendicular to the axis of rotation. Then the *angular position* of that point is the angle θ from a reference axis (typically the positive x-axis) to the vector $\mathbf{r}_\perp(t)$ in a known rotation sense (typically given by the right-hand rule).

- Angular velocity : the angular velocity ω is the rate at which the angular position θ changes with respect to time t:

$$\omega = \frac{d\theta}{dt}$$

The angular velocity is represented in Figure 1 by a vector **Ω** pointing along the axis of rotation with magnitude ω and sense determined by the direction of rotation as given by the right-hand rule.

- Angular acceleration : the magnitude of the angular acceleration α is the rate at which the angular velocity ω changes with respect to time t:

$$\alpha = \frac{d\omega}{dt}$$

The equations of translational kinematics can easily be extended to planar rotational kinematics for constant angular acceleration with simple variable exchanges:

$$\omega_f = \omega_i + \alpha t$$
$$\theta_f - \theta_i = \omega_i t + \tfrac{1}{2}\alpha t^2$$
$$\theta_f - \theta_i = \tfrac{1}{2}(\omega_f + \omega_i)t$$
$$\omega_f^2 = \omega_i^2 + 2\alpha(\theta_f - \theta_i)$$

Here θ_i and θ_f are, respectively, the initial and final angular positions, ω_i and ω_f are, respectively, the initial and final angular velocities, and α is the constant angular acceleration. Although position in space and velocity in space are both true vectors (in terms of their properties under rotation), as is angular velocity, angle itself is not a true vector.

Point Trajectories in Body Moving in Three Dimensions

Important formulas in kinematics define the velocity and acceleration of points in a moving body as they trace trajectories in three-dimensional space. This is particularly important for the center of mass of a body, which is used to derive equations of motion using either Newton's second law or Lagrange's equations.

Position

In order to define these formulas, the movement of a component B of a mechanical system is defined by the set of rotations [A(t)] and translations d(t) assembled into the homogeneous transformation [T(t)]=[A(t), d(t)]. If p is the coordinates of a point P in B measured in the moving reference frame M, then the trajectory of this point traced in F is given by:

$$\mathbf{P}(t) = [T(t)]\mathbf{p} = \begin{Bmatrix} \mathbf{P} \\ 1 \end{Bmatrix} = \begin{bmatrix} A(t) & \mathbf{d}(t) \\ 0 & 1 \end{bmatrix} \begin{Bmatrix} \mathbf{p} \\ 1 \end{Bmatrix}.$$

This notation does not distinguish between **P** = (X, Y, Z, 1), and **P** = (X, Y, Z), which is hopefully clear in context.

This equation for the trajectory of P can be inverted to compute the coordinate vector p in M as:

$$\mathbf{p} = [T(t)]^{-1}\mathbf{P}(t) = \begin{Bmatrix} \mathbf{p} \\ 1 \end{Bmatrix} = \begin{bmatrix} A(t)^T & -A(t)^T\mathbf{d}(t) \\ 0 & 1 \end{bmatrix} \begin{Bmatrix} \mathbf{P}(t) \\ 1 \end{Bmatrix}.$$

This expression uses the fact that the transpose of a rotation matrix is also its inverse, that is:

$$[A(t)]^T[A(t)] = I.$$

Velocity

The velocity of the point *P* along its trajectory **P**(t) is obtained as the time derivative of this position vector,

$$\mathbf{V}_P = [\dot{T}(t)]\mathbf{p} = \begin{Bmatrix} \mathbf{V}_P \\ 0 \end{Bmatrix} = \begin{bmatrix} A(t) & \mathbf{d}(t) \\ 0 & 1 \end{bmatrix} \begin{Bmatrix} \mathbf{p} \\ 1 \end{Bmatrix} = \begin{bmatrix} \dot{A}(t) & \dot{\mathbf{d}}(t) \\ 0 & 0 \end{bmatrix} \begin{Bmatrix} \mathbf{p} \\ 1 \end{Bmatrix}.$$

The dot denotes the derivative with respect to time; because **p** is constant, its derivative is zero.

The matrix [S] is given by:

$$[S] = \begin{bmatrix} \Omega & -\Omega\mathbf{d} + \dot{\mathbf{d}} \\ 0 & 0 \end{bmatrix}$$

where

$$[\Omega] = \dot{A}A^T,$$

is the angular velocity matrix.

Multiplying by the operator [S], the formula for the velocity $\mathbf{V}_P$ takes the form:

$$\mathbf{V}_P = [\Omega](\mathbf{P} - \mathbf{d}) + \dot{\mathbf{d}} = \omega \times \mathbf{R}_{P/O} + \mathbf{V}_O,$$

where the vector ω is the angular velocity vector obtained from the components of the matrix [Ω]; the vector

$$\mathbf{R}_{P/O} = \mathbf{P} - \mathbf{d},$$

is the position of *P* relative to the origin *O* of the moving frame *M*; and

$$\mathbf{V}_O = \dot{\mathbf{d}},$$

is the velocity of the origin *O*.

Acceleration

The acceleration of a point *P* in a moving body *B* is obtained as the time derivative of its velocity vector:

$$\mathbf{A}_P = \frac{d}{dt}\mathbf{V}_P = \frac{d}{dt}\left([S]\mathbf{P}\right) = [\dot{S}]\mathbf{P} + [S]\dot{\mathbf{P}} = [\dot{S}]\mathbf{P} + [S][S]\mathbf{P}.$$

This equation can be expanded firstly by computing

$$[\dot{S}] = \begin{bmatrix} \dot{\Omega} & -\dot{\Omega}\mathbf{d} - \Omega\dot{\mathbf{d}} + \ddot{\mathbf{d}} \\ 0 & 0 \end{bmatrix} = \begin{bmatrix} \dot{\Omega} & -\dot{\Omega}\mathbf{d} - \Omega\mathbf{V}_O + \mathbf{A}_O \\ 0 & 0 \end{bmatrix}$$

and

$$[S]^2 = \begin{bmatrix} \Omega & -\Omega\mathbf{d} + \mathbf{V}_O \\ 0 & 0 \end{bmatrix}^2 = \begin{bmatrix} \Omega^2 & -\Omega^2\mathbf{d} + \Omega\mathbf{V}_O \\ 0 & 0 \end{bmatrix}.$$

The formula for the acceleration $\mathbf{A}_P$ can now be obtained as:

$$\mathbf{A}_P = \dot{\Omega}(\mathbf{P} - \mathbf{d}) + \mathbf{A}_O + \Omega^2(\mathbf{P} - \mathbf{d}),$$

or

$$\mathbf{A}_P = \alpha \times \mathbf{R}_{P/O} + \omega \times \omega \times \mathbf{R}_{P/O} + \mathbf{A}_O,$$

where α is the angular acceleration vector obtained from the derivative of the angular velocity matrix;

$$\mathbf{R}_{P/O} = \mathbf{P} - \mathbf{d},$$

is the relative position vector (the position of *P* relative to the origin *O* of the moving frame *M*); and

$$\mathbf{A}_O = \ddot{\mathbf{d}}$$

is the acceleration of the origin of the moving frame *M*.

Kinematic Constraints

Kinematic constraints are constraints on the movement of components of a mechanical system. Kinematic constraints can be considered to have two basic forms, (i) constraints that arise from hinges, sliders and cam joints that define the construction of the system, called holonomic constraints, and (ii) constraints imposed on the velocity of the system such as the knife-edge constraint of ice-skates on a flat plane, or rolling without slipping of a disc or sphere in contact with a plane, which are called non-holonomic constraints. The following are some common examples.

Kinematic Coupling

A kinematic coupling exactly constrains all 6 degrees of freedom.

Rolling without Slipping

An object that rolls against a surface without slipping obeys the condition that the velocity of its center of mass is equal to the cross product of its angular velocity with a vector from the point of contact to the center of mass:

$$\mathbf{v}_G(t) = \Omega \times \mathbf{r}_{G/O}.$$

For the case of an object that does not tip or turn, this reduces to $v = r\omega$.

Inextensible Cord

This is the case where bodies are connected by an idealized cord that remains in tension and cannot change length. The constraint is that the sum of lengths of all segments of the cord is the total length, and accordingly the time derivative of this sum is zero. A dynamic problem of this type is the pendulum. Another example is a drum turned by the pull of gravity upon a falling weight attached to the rim by the inextensible cord. An *equilibrium* problem (i.e. not kinematic) of this type is the catenary.

Kinematic Pairs

Reuleaux called the ideal connections between components that form a machine kinematic pairs. He distinguished between higher pairs which were said to have line contact between the two links and lower pairs that have area contact between the links. J. Phillips shows that there are many ways to construct pairs that do not fit this simple classification.

Lower Pair

A lower pair is an ideal joint, or holonomic constraint, that maintains contact between a point, line or plane in a moving solid (three-dimensional) body to a corresponding point line or plane in the fixed solid body. There are the following cases:

- A revolute pair, or hinged joint, requires a line, or axis, in the moving body to remain co-linear with a line in the fixed body, and a plane perpendicular to this line in the moving body maintain contact with a similar perpendicular plane in the fixed body. This imposes five constraints on the relative movement of the links, which therefore has one degree of freedom, which is pure rotation about the axis of the hinge.

- A prismatic joint, or slider, requires that a line, or axis, in the moving body remain co-linear with a line in the fixed body, and a plane parallel to this line in the moving body maintain contact with a similar parallel plane in the fixed body. This imposes five constraints on the relative movement of the links, which therefore has one degree of freedom. This degree of freedom is the distance of the slide along the line.
- A cylindrical joint requires that a line, or axis, in the moving body remain co-linear with a line in the fixed body. It is a combination of a revolute joint and a sliding joint. This joint has two degrees of freedom. The position of the moving body is defined by both the rotation about and slide along the axis.
- A spherical joint, or ball joint, requires that a point in the moving body maintain contact with a point in the fixed body. This joint has three degrees of freedom.
- A planar joint requires that a plane in the moving body maintain contact with a plane in fixed body. This joint has three degrees of freedom.

Higher Pairs

Generally speaking, a higher pair is a constraint that requires a curve or surface in the moving body to maintain contact with a curve or surface in the fixed body. For example, the contact between a cam and its follower is a higher pair called a *cam joint*. Similarly, the contact between the involute curves that form the meshing teeth of two gears are cam joints.

Kinematic Chains

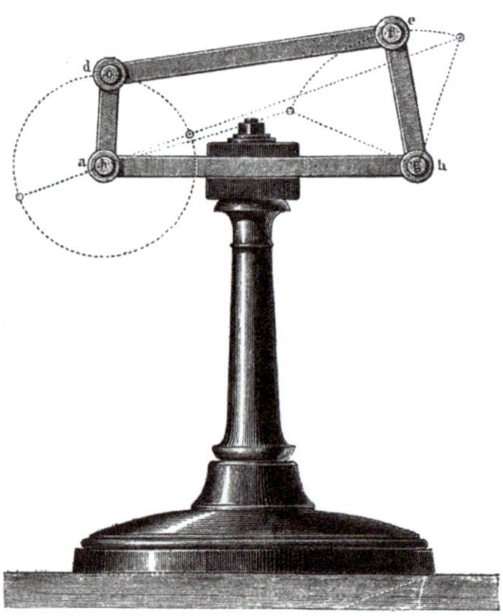

The_Kinematics_of_Machinery Kinematics of Machinery, 1876

Rigid bodies ("links") connected by kinematic pairs ("joints") are known as *kinematic chains*. Mechanisms and robots are examples of kinematic chains. The degree of freedom of a kinematic chain is computed from the number of links and the number and type of joints using the mobility formula. This formula can also be used to enumerate the topologies of kinematic chains that have a given degree of freedom, which is known as *type synthesis* in machine design.

Examples

The planar one degree-of-freedom linkages assembled from N links and j hinged or sliding joints are:

- N=2, j=1 : a two-bar linkage that is the lever;
- N=4, j=4 : the four-bar linkage;
- N=6, j=7 : a six-bar linkage. This must have two links ("ternary links") that support three joints. There are two distinct topologies that depend on how the two ternary linkages are connected. In the Watt topology, the two ternary links have a common joint; in the Stephenson topology, the two ternary links do not have a common joint and are connected by binary links.
- N=8, j=10 : eight-bar linkage with 16 different topologies;
- N=10, j=13 : ten-bar linkage with 230 different topologies;
- N=12, j=16 : twelve-bar linkage with 6,856 topologies.

For larger chains and their linkage topologies, see R. P. Sunkari and L. C. Schmidt, "Structural synthesis of planar kinematic chains by adapting a Mckay-type algorithm", *Mechanism and Machine Theory* #41, pp. 1021–1030 (2006).

Technical Drawing

Technical drawing, drafting or draughting, is the act and discipline of composing drawings that visually communicate how something functions or is to be constructed.

Technical drawing is essential for communicating ideas in industry and engineering. To make the drawings easier to understand, people use familiar symbols, perspectives, units of measurement, notation systems, visual styles, and page layout. Together, such conventions constitute a visual language, and help to ensure that the drawing is unambiguous and relatively easy to understand. These drafting conventions are condensed into internationally accepted standards and specifications that transcend the barrier of language making technical drawings a universal means of communicating complex mechanical concepts.

Drafter at work

This need for precise communication in the preparation of a functional document distinguishes technical drawing from the expressive drawing of the visual arts. Artistic drawings are subjectively interpreted; their meanings are multiply determined. Technical drawings are understood to have one intended meaning.

Bundesarchiv, B 145 Bild-F038800-0010
Foto: Schaack, Lothar | 1. März 1973

Copying technical drawings in 1973

A *drafter, draftsperson*, or *draughtsman* is a person who makes a drawing (technical or expressive). A professional drafter who makes technical drawings is sometimes called a *drafting technician*. Professional drafting is a desirable and necessary function in the design and manufacture of complex mechanical components and machines. Professional draftspersons bridge the gap between engineers and manufacturers, and contribute experience and technical expertise to the design process.

Methods

Manual or by Instrument

The basic drafting procedure is to place a piece of paper (or other material) on a smooth surface with right-angle corners and straight sides—typically a drawing board. A sliding straightedge known as a *T-square* is then placed on one of the sides, allowing it to be slid across the side of the table, and over the surface of the paper.

A drafting table

"Parallel lines" can be drawn simply by moving the T-square and running a pencil or technical pen along the T-square's edge. The T-square is used to hold other devices such as set squares or triangles. In this case the drafter places one or more triangles of known angles on the T-square—which is itself at right angles to the edge of the table—and can then draw lines at any chosen angle to others on the page. Modern drafting tables come equipped with a drafting machine that is supported on both sides of the table to slide over a large piece of paper. Because it is secured on both sides, lines drawn along the edge are guaranteed to be parallel.

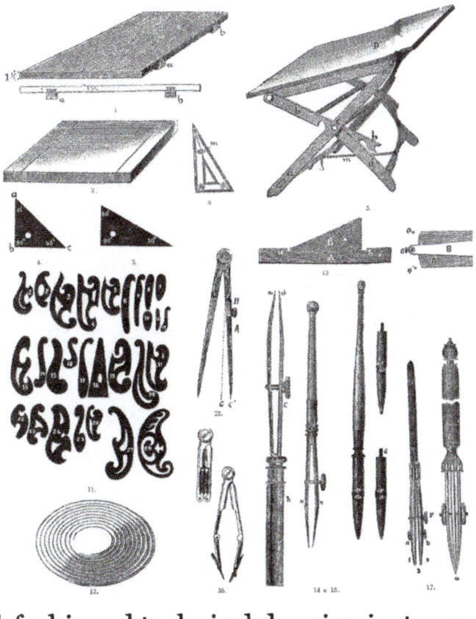

Old-fashioned technical drawing instruments

In addition, the drafter uses several technical drawing tools to draw curves and circles. Primary among these are the compasses, used for drawing simple arcs and circles, and the French curve,for drawing curves. A spline is a rubber coated articulated metal that can be manually bent to most curves.

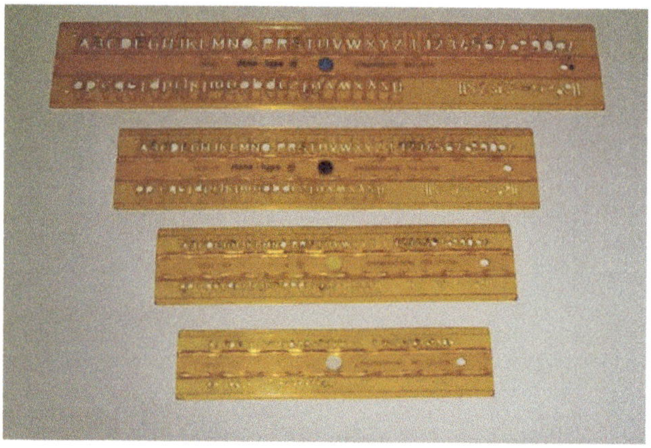

Stencils for lettering technical drawings to DIN standards

Drafting templates assist the drafter with creating recurring objects in a drawing without having to reproduce the object from scratch every time. This is especially useful when using common symbols; i.e. in the context of stagecraft, a lighting designer will draw from the USITT standard library of lighting fixture symbols to indicate the position of a common fixture across multiple positions. Templates are sold commercially by a number of vendors, usually customized to a specific task, but it is also not uncommon for a drafter to create his own templates.

This basic drafting system requires an accurate table and constant attention to the positioning of the tools. A common error is to allow the triangles to push the top of the T-square down slightly, thereby throwing off all angles. Even tasks as simple as drawing two angled lines meeting at a point require a number of moves of the T-square and triangles, and in general drafting can be a time-consuming process.

A solution to these problems was the introduction of the mechanical "drafting machine", an application of the pantograph (sometimes referred to incorrectly as a "pentagraph" in these situations) which allowed the drafter to have an accurate right angle at any point on the page quite quickly. These machines often included the ability to change the angle, thereby removing the need for the triangles as well.

In addition to the mastery of the mechanics of drawing lines, arcs and circles (and text) onto a piece of paper—with respect to the detailing of physical objects—the drafting effort requires a thorough understanding of geometry, trigonometry and spatial comprehension, and in all cases demands precision and accuracy, and attention to detail of high order.

Although drafting is sometimes accomplished by a project engineer, architect—or even by shop personnel such as a machinist—skilled drafters (and/or designers) usually accomplish the task and are always in demand to some level.

Computer Aided Drafting/Design

Today, the mechanics of the drafting task have largely been automated and accelerated through the use of computer-aided design systems (CAD).

There are two types of computer-aided design systems used for the production of technical drawings" two dimensions ("2D") and three dimensions ("3D").

2D CAD systems such as AutoCAD or MicroStation replace the paper drawing discipline. The lines, circles, arcs and curves are created within the software. It is down to the technical drawing skill of the user to produce the drawing. There is still much scope for error in the drawing when producing first and third angle orthographic projections, auxiliary projections and cross sections. A 2D CAD system is merely an electronic drawing board. Its greatest strength over direct to paper technical drawing is in the making of revisions. Whereas in a conventional hand drawn technical drawing, if a mistake is found, or a modification is required, a new drawing must be made from scratch, the 2D CAD system allows a copy of the original to be modified, saving considerable time. 2D CAD systems can be used to create plans for large projects such as buildings and aircraft but provide no way to check the various components will fit together.

View of a CAD model of a four-cylinder, inline crankshaft with pistons

A 3D CAD system (such as KeyCreator, Autodesk Inventor, or SolidWorks) first produces the geometry of the part; the technical drawing comes from user defined views of that geometry. Any orthographic, projected or sectioned view is created by the software. There is no scope for error in the production of these views. The main scope for error comes in setting the parameter of first or third angle projection, and displaying the relevant symbol on the technical drawing. 3D CAD allows individual parts to be assembled together to represent the final product. Buildings, Aircraft, ships, and cars are modeled, assembled, and checked in 3D before technical drawings are released for manufacture.

Both 2D and 3D CAD systems can be used to produce technical drawings for any discipline. The various disciplines (electrical, electronic, pneumatic, hydraulic, etc.) have industry recognized symbols to represent common components.

BS and ISO produce standards to show recommended practices but it is up to individuals to produce the drawings. There is no definitive standard for layout or style. The only standard across engineering workshop drawings is in the creation of orthographic projections and cross section views.

Drafting can represent two dimensions ("2D") and three dimensions ("3D") although the representation itself is always created in 2D (cf. Architectural model). Drafting is the integral communication of technical or engineering drawings and is the industrial arts sub-discipline that underlies all involved technical endeavors.

In representing complex, three-dimensional objects in two-dimensional drawings, the objects can be described by at least one view plus material thickness note, 2, 3 or as many views and sections that are required to show all features of object.

Applications for Technical Drawing

Architecture

To plan a renovation, this architect takes measurements, which he later enters into his computer-aided design software.

The art and design that goes into making buildings is known as "architecture". To communicate all aspects of the shape or design, detail drawings are used. In this field, the term *plan* is often used when referring to the full section view of these drawings as viewed from three feet above finished floor to show the locations of doorways, windows, stairwells, etc. Architectural drawings describe and document an architect's design.

Engineering

Engineering can be a very broad term. It stems from the Latin *ingenerare*, meaning "to create". Because this could apply to everything that humans create, it is given a narrower definition in the context of technical drawing. Engineering drawings generally deal with mechanical engineered items, such as manufactured parts and equipment.

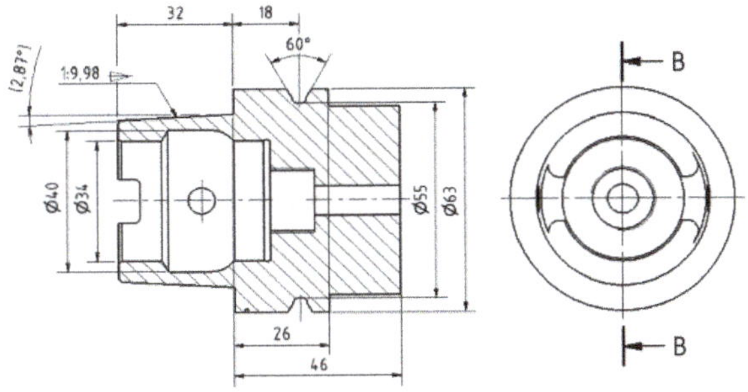

Engineering drawing of a machine tool part

Engineering drawings are usually created in accordance with standardized conventions for layout, nomenclature, interpretation, appearance (such as typefaces and line styles), size, etc.

Its purpose is to accurately and unambiguously capture all the geometric features of a product or a component. The end goal of an engineering drawing is to convey all the required information that will allow a manufacturer to produce that component.

Related Fields

Technical Illustration

Illustration of a drum set

Technical illustration is the use of illustration to visually communicate information of a technical nature. Technical illustrations can be component technical drawings or diagrams. The aim of technical illustration is "to generate expressive images that effectively convey certain information via the visual channel to the human observer".

The main purpose of technical illustration is to describe or explain these items to a more or less nontechnical audience. The visual image should be accurate in terms of dimensions and proportions, and should provide "an overall impression of what an object is or does, to enhance the viewer's interest and understanding".

According to Viola (2005) "illustrative techniques are often designed in a way that even a person with no technical understanding clearly understands the piece of art. The use of varying line widths to emphasize mass, proximity, and scale helped to make a simple line drawing more understandable to the lay person. Cross hatching, stippling, and other low abstraction techniques gave greater depth and dimension to the subject matter".

Cutaway Drawing

Cutaway drawing of the Nash 600, an American car of the 1940s

A *cutaway drawing* is a technical illustration, in which part of the surface of a three-dimensional model is removed in order to show some of the model's interior in relation to its exterior.

The purpose of a cutaway drawing is to "allow the viewer to have a look into an otherwise solid opaque object. Instead of letting the inner object shine through the surrounding surface, parts of outside object are simply removed. This produces a visual appearance as if someone had cutout a piece of the object or sliced it into parts. Cutaway illustrations avoid ambiguities with respect to spatial ordering, provide a sharp contrast between foreground and background objects, and facilitate a good understanding of spatial ordering".

Technical Drawings

Types of Technical Drawings

The two types of technical drawings are based on graphical projection. This is used to create an image of a three-dimensional object onto a two-dimensional surface.

Two-dimensional Representation

Two-dimensional representation uses orthographic projection to create an image where only two of the three dimensions of the object are seen.

Three-dimensional Representation

In a three-dimensional representation, also referred to as a pictorial, all three dimensions of an object are visible.

Views

Multiview

Multiview is a type of orthographic projection. There are two conventions for using multiview, first-angle and third-angle. In both cases, the front or main side of the object is the same. First-angle is drawing the object sides based on where they land. Example, looking at the front side, rotate the object 90 degrees to the right. What is seen will be drawn to the right of the front side. Third-angle is drawing the object sides based on where they are. Example, looking at the front side, rotate the object 90 degrees to the right. What is seen is actually the left side of the object and will be drawn to the left of the front side

Section

While multiview relates to external surfaces of an object, section views show an imaginary plane cut through an object. This is often useful to show voids in an object.

Auxiliary

Auxiliary views utilize an additional projection plane other than the common planes in a multiview. Since the features of an object need to show the true shape and size of the object, the projection plane must be parallel to the object surface. Therefore, any surface that is not in line with the three major axis needs its own projection plane to show the features correctly.

Pattern

Patterns, sometimes called developments, show the size and shape of a flat piece of material needed for later bending or folding into a three dimensional shape.

Exploded

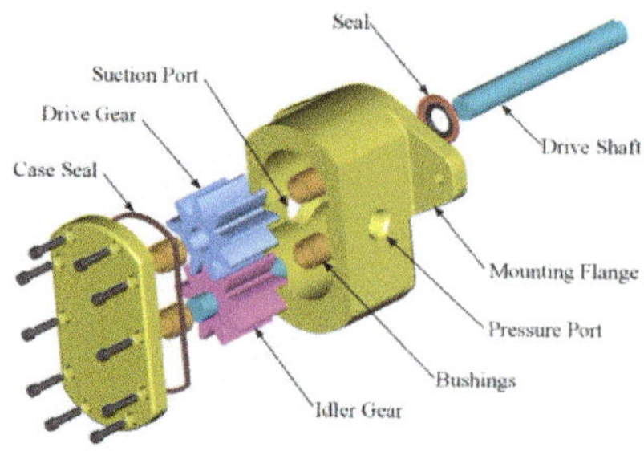

Exploded-view drawing of a gear pump

An *exploded-view drawing* is a technical drawing of an object that shows the relationship or order of assembly of the various parts. It shows the components of an object slightly separated by distance, or suspended in surrounding space in the case of a three-dimensional exploded diagram. An object is represented as if there had been a small controlled explosion emanating from the middle of the object, causing the object's parts to be separated relative distances away from their original locations.

An exploded view drawing (EVD) can show the intended assembly of mechanical or other parts. In mechanical systems usually the component closest to the center is assembled first, or is the main part in which the other parts get assembled. This drawing can also help to represent disassembly of parts, where the parts on the outside normally get removed first.

Standards and Conventions

Basic Drafting Paper Sizes

There have been many standard sizes of paper at different times and in different countries, but today most of the world uses the international standard (A4 and its siblings). North America uses its own sizes.

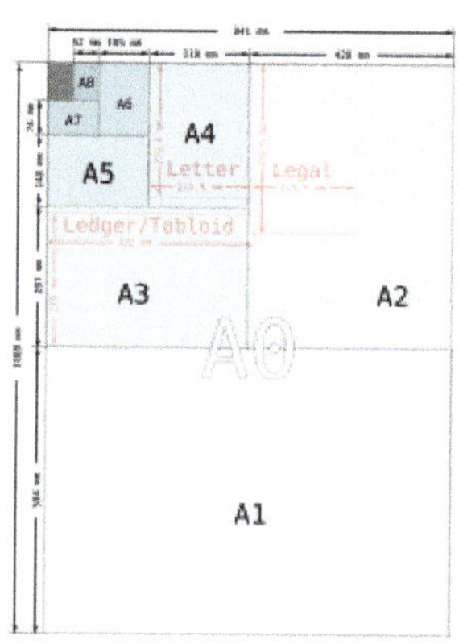

ISO "A series" papers sizes used in most countries of the world

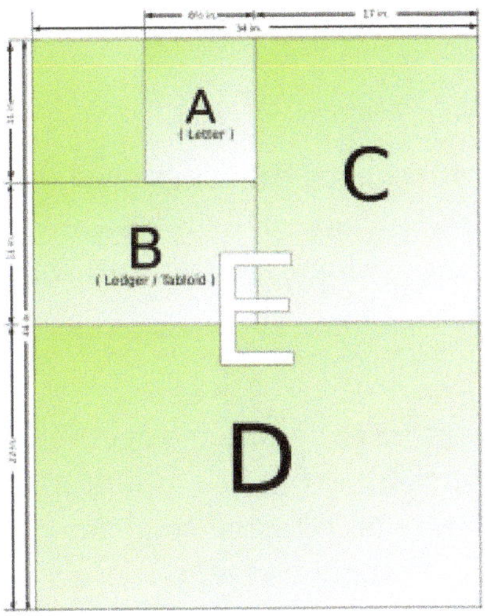

ANSI paper sizes used in North America

Patent Drawing

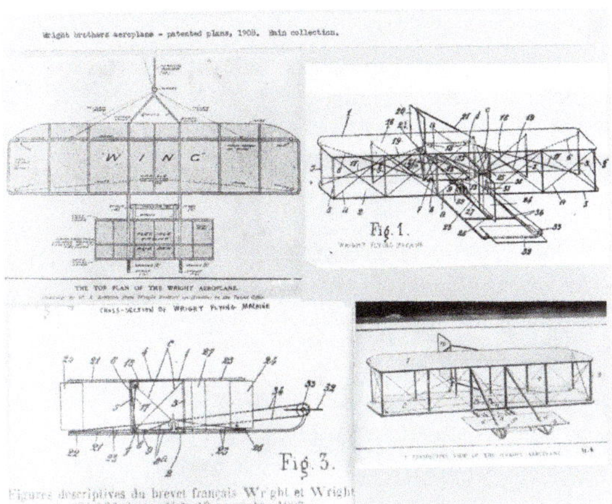

The Wright brothers' patent drawings for their aeroplane (1908)

The applicant for a patent will be required by law to furnish a drawing of the invention if or when the nature of the case requires a drawing to understand the invention with the job. This drawing must be filed with the application. This includes practically all inventions except compositions of matter or processes, but a drawing may also be useful in the case of many processes.

The drawing must show every feature of the invention specified in the claims, and is required by the patent office rules to be in a particular form. The Office specifies the size of the sheet on which the drawing is made, the type of paper, the margins, and other details relating to the making of the drawing. The reason for specifying the standards in detail is that the drawings are printed and published in a uniform style when the patent issues, and the drawings must also be such that they can be readily understood by persons using the patent descriptions.

Sets of Technical Drawings

Working Drawings for Production

Working drawings are the set of technical drawings used during the manufacturing phase of a product. In architecture, these include civil drawings, architectural drawings, structural drawings, mechanical systems drawings, electrical drawings, and plumbing drawings.

Assembly Drawings

Assembly drawings show how different parts go together, identify those parts by number, and have a parts list, often referred to as a bill of materials. In a technical service manual, this type of drawing may be referred to as an exploded view drawing or diagram. These parts may be used in engineering.

As-fitted Drawings

Also called As-Built drawings, or As-made drawings. As-fitted drawings represent a record of the completed works, literally 'as fitted'. These are based upon the working drawings and updated to reflect any changes or alterations undertaken during construction or manufacture.

References

- Edmund Taylor Whittaker (1904). A Treatise on the Analytical Dynamics of Particles and Rigid Bodies. Cambridge University Press. Chapter 1. ISBN 0-521-35883-3.
- Russell C. Hibbeler (2009). "Kinematics and kinetics of a particle". Engineering Mechanics: Dynamics (12th ed.). Prentice Hall. p. 298. ISBN 0-13-607791-9.
- Ahmed A. Shabana (2003). "Reference kinematics". Dynamics of Multibody Systems (2nd ed.). Cambridge University Press. ISBN 978-0-521-54411-5.
- P. P. Teodorescu (2007). "Kinematics". Mechanical Systems, Classical Models: Particle Mechanics. Springer. p. 287. ISBN 1-4020-5441-6. .
- Paul, Richard (1981). Robot manipulators: mathematics, programming, and control : the computer control of robot manipulators. MIT Press, Cambridge, MA. ISBN 978-0-262-16082-7.
- William Thomson Kelvin & Peter Guthrie Tait (1894). Elements of Natural Philosophy. Cambridge University Press. p. 4. ISBN 1-57392-984-0.
- M. Fogiel (1980). "Problem 17-11". The Mechanics Problem Solver. Research & Education Association. p. 613. ISBN 0-87891-519-2.
- Morris Kline (1990). Mathematical Thought from Ancient to Modern Times. Oxford University Press. p. 472. ISBN 0-19-506136-5.
- Phillips, Jack (2007). Freedom in Machinery, Volumes 1-2 (reprint ed.). Cambridge University Press. ISBN 978-0-521-67331-0.
- Tsai, Lung-Wen (2001). Mechanism design:enumeration of kinematic structures according to function (llustrated ed.). CRC Press. p. 121. ISBN 978-0-8493-0901-4.

- Goetsch, David L.; Chalk, William S.; Nelson, John A. (2000). Technical Drawing. Delmar Technical Graphics Series (Fourth ed.). Albany: Delmar Learning. p. 3. ISBN 978-0-7668-0531-6. OCLC 39756434.
- Jefferis, Alan; Madsen, David (2005), Architectural Drafting and Design (5th ed.), Clifton Park, NY: Delmar Cengage Learning, ISBN 1-4018-6715-4
- Lieu, Dennis K; Sorby, Sheryl (2009), Visualization, Modeling, and Graphics for Engineering Design (1st ed.), Clifton Park, NY: Delmar Cengage Learning, ISBN 1-4018-4249-6, p. 1-2
- Michael E. Brumbach, Jeffrey A. Clade (2003). Industrial Maintenance. Cengage Learning, 2003 ISBN 0-7668-2695-3, p.65
- Ralph W. Liebing (1999). Architectural working drawings. John Wiley and Sons, 1999. ISBN 0-471-34876-7.

4

Technologies Related to Manufacturing

Power looms are looms that can be controlled mechanically whereas automation is the use of control systems for using equipment such as machinery, boilers and steering. Some other technologies explained in this section are conveyor system, conveyor belt, computer-aided manufacturing, computer-integrated manufacturing, pilot plant etc. The aim of this chapter is to explore all the technologies related to manufacturing.

Power Loom

A Northrop loom manufactured by Draper Corporation in the textile museum, Lowell, Massachusetts.

A power loom is a mechanised loom powered by a line shaft, and was one of the key developments in the industrialization of weaving during the early Industrial Revolution. The first power loom was designed in 1784 by Edmund Cartwright and first built in 1785. It was refined over the next 47 years until a design by Kenworthy and Bullough made the operation completely automatic.

By 1850 there were 260,000 in operation in England. Fifty years later came the Northrop Loom that would replenish the shuttle when it was empty and this replaced the Lancashire loom.

Shuttle Looms

The huge components of the loom are the warp beam, heddles, harnesses, shuttle, reed and takeup roll. In the loom, yarn processing includes shedding, picking, battening and taking-up operations.

- *Shedding.* Shedding is the raising of the warp yarns to form a loop through which the filling yarn, carried by the shuttle, can be inserted. The shed is the vertical space between the raised and unraised warp yarns. On the modern loom, simple and intricate shedding operations are performed automatically by the heddle or heald frame, also known as a harness. This is a rectangular frame to which a series of wires, called heddles or healds, are attached. The yarns are passed through the eye holes of the heddles, which hang vertically from the harnesses. The weave pattern determines which harness controls which warp yarns, and the number of harnesses used depends on the complexity of the weave. Two common methods of controlling the heddles are dobbies and a Jacquard Head.

Shuttle with pirn

- *Picking.* As the harnesses raise the heddles or healds, which raise the warp yarns, the shed is created. The filling yarn is inserted through the shed by a small carrier device called a shuttle. The shuttle is normally pointed at each end to allow passage through the shed. In a traditional shuttle loom, the filling yarn is wound onto a quill, which in turn is mounted in the shuttle. The filling yarn emerges through a hole in the shuttle as it moves across the loom. A single crossing of the shuttle from one side of the loom to the other is known as a pick. As the shuttle moves back and forth across the shed, it weaves an edge, or selvage, on e-ach side of the fabric to prevent the fabric from ravelling.

- *Battening.* As the shuttle moves across the loom laying down the fill yarn, it also passes through openings in another frame called a reed (which resembles a comb). With each picking operation, the reed presses or battens each filling yarn against the portion of the fabric that has already been formed. The point where the fabric is formed is called the fell. Conventional shuttle looms can operate at speeds of about 150 to 160 picks per minute.

With each weaving operation, the newly constructed fabric must be wound on a cloth beam. This process is called taking up. At the same time, the warp yarns must be let off or released from the warp beams. To become fully automatic, a loom needs a filling stop motion which will brake the loom, if the weft thread breaks.

Operation

Operation of weaving in a textile mill is undertaken by a specially trained operator known as a weaver. Weavers are expected to uphold high industry standards, and are tasked with monitoring anywhere from ten, to as many as thirty separate looms at any one time. During their operating shift, weavers will first utilize a wax pencil or crayon to sign their initials onto the cloth to mark a shift change, and then walk along the cloth side (front) of the looms they tend, gently touching the fabric as it comes from the reed. This is done to feel for any broken "picks" or filler thread. Should broken picks be detected, the weaver will disable the machine and undertake to correct the error, typically by replacing the bobbin of filler thread in as little time as possible. They are trained that, ideally, no machine should stop working for more than one minute, with faster turn around times being preferred.

Once the weaver has made their circuit of the front of the machines, they will then circle around to the back. At this point they will gently stroke their hand over the raised metal "tells" on the back of the machine. These tells, located over a special metal circuit, are held up by the tension of the thread coming from the warp. Should the warp thread be broken, the tells will drop and cause the machine to stop working. However, it is possible for them to become stuck in the upward position, and by doing so create problems in the weaving. By gently touching the tells, then, it is possible for the weaver to find tells which have become stuck in the up position, and correct the error. As with pick breaks, the weavers are trained to keep the machines running as much as possible; with speedy knot tying and correction being stressed. In this situation, they are expected to take less than a minute, with the mean ideal being ten to thirty seconds, to correct a break. The weaver also watches for warps that are about to run out, or problems in the warp itself which were not detected in the slashing process. Typically, weavers can expect to make several dozen circuits of their machines a night, with most of their time spent ensuring the quality of the cloth and the company standards of production.

History

Edmund Cartwright patented a power loom in 1785. This used water as power instead of human power which sped up the weaving process. Weavers were able to use all the thread that spinners could produce. It was to be forty years before his ideas were modified into a reliable automatic loom. Cartwright was not the first man to design an automatic loom, this had been done in 1678 by M. de Gennes in Paris, and again by Vaucanson In 1745, but these never developed and were forgotten. Those designs preceded John Kay's invention of the flying shuttle and they passed the shuttle through the shed using levers.

A loom from the 1890s with a dobby head. Illustration from the Textile Mercury.

It was not a commercially successful machine. His ideas were licensed first by Grimshaw, of Manchester who built a small steam-powered weaving factory in Manchester in 1790. The looms had to be stopped to dress the warp, but the factory burnt down before anything could be learned.

Series of Initial Inventors

A series of inventors incrementally improved all aspects of the three principle processes and the ancillary processes.

- Grimshaw 1790 Manchester- dressing the warp
- Austin 1789, 1790 -dressing the warp, 200 looms produced for Monteith of Pollockshaws 1800
- Thomas Johnson, 1803, Bredbury- dressing frame: Factory for 200 Steam Looms on Manchester 1806, and two factories at Stockport 1809. One at Westhoughton, Lancashire 1809.
- William Radcliffe of Stockport 1802- improved take up mechanism
- John Todd of Burnley 1803- a heald roller and new shedding arrangements, the healds were corded to treadles actuated by cams on the second shaft.
- William Horrocks of Stockport 1803- The frame was still wooden but the lathe was pendant from the frame and operated by cams on the first shaft, the shedding was operated by cams on the second shaft, the take up motion was copied from Radcliffe.

- Peter Marsland 1806- improvements to the lathe motion to counteract poor picking
- William Cotton 1810- improvements to the letting off motion
- William Horrocks 1813 -Horrocks Loom Modifications to the lathe motion- improving on Marsland
- Peter Ewart 1813 -a use of pneumatics
- Joseph and Peter Taylor 1815 -double beat foot lathe for heavy cloths
- Paul Moody 1815- produces the first power loom in North America. Exporting a UK loom would have been illegal.
- John Capron and Sons 1820- installed the first power looms for woolens in North America at Uxbridge, Massachusetts.
- William Horrocks 1821 -a system to wet the warp and weft during use, improving the effectiveness of the sizing
- Richard Roberts 1830, Roberts Loom, These improvements were a geared take up wheel and tappets to operate multiple heddles
- Stanford, Pritchard and Wilkinson- patented a method to stop on the break of weft or warp. It was not used.
- William Dickinson of Blackburn Blackburn Loom the modern overpick loom

Further Useful Improvements

There now appear a series of useful improvements that are contained in patents for useless devices

- Hornby, Kenworthy and Bullough of Blackburn 1834- the vibrating or fly reed
- John Ramsbottom and Richard Holt of Todmorden 1834- a new automatic weft stopping motion
- James Bullough of Blackburn 1835- improved automatic weft stopping motion and taking up and letting off arrangements
- Andrew Parkinson 1836- improved stretcher (temple).
- William Kenworthy and James Bullough 1841- trough and roller temple (became the standard), A simple stop-motion.

At this point the loom has become automatic except for refilling weft pirns. The Cartwight loom weaver could work one loom at 120-130 picks per minute- with a Kenworthy and Bullough's Lancashire Loom, a weaver can run up to six looms working at 220-260 picks per minute- thus giving 12 times more through put.

Looms and the Manchester Context

The development of the power loom in and around Manchester was not a coincidence. Manchester has been a centre for Fustians by 1620 and acted as a hub for other Lancashire towns, so developing a communication network with them. It was an established point of export using the meandering River Mersey, and by 1800 it had a thriving canal network, with links to the Ashton Canal, Rochdale Canal the Peak Forest Canal and Manchester Bolton & Bury Canal. The fustian trade gave the towns a skilled workforce that was used to the complicated Dutch looms, and was perhaps accustomed to industrial discipline. While Manchester became a spinning town, the towns around were weaving towns producing cloth by the *putting out* system. The business was dominated by a few families who had the capital needed for the investment in new mills, and buy hundreds of looms. The mills were built along the new canals so immediately had access to their markets. Spinning developed first and, until 1830, the handloom was still more important economically than the power loom when the roles reversed. Because of the economic growth of Manchester, a new industry of precision machine tool engineering was born and here were the skills needed to build the precision mechanisms of a loom.

Adoption

Number of Looms in UK					
Year	1803	1820	1829	1833	1857
Looms	2,400	14,650	55,500	100,000	250,000

Draper's strategy was to standardise on a couple of models which it mass-produced. The lighter E-model of 1909 was joined in the 1930 by the heavier X-model. Continuous fibre machines, say for rayon, which was more break-prone, needed a specialist loom. This was provided by the purchase of the Stafford Loom Co. in 1932, and using their patents a third loom the XD, was added to the range. Because of their mass production techniques they were reluctant and slow to retool for new technologies such as shuttleless looms.

Decline and Reinvention

Originally, power looms used a shuttle to throw the weft across, but in 1927 the faster and more efficient shuttleless loom came into use. Sulzer Brothers, a Swiss company had the exclusive rights to shuttleless looms in 1942, and licensed the American production to Warner & Swasey. Draper licensed the slower rapier loom. Today, advances in technology have produced a variety of looms designed to maximise production for specific types of material. The most common of these are Sulzer shuttleless weaving machines, rapier looms, air-jet looms and water-jet looms.

Social and Economic Implications

The power loom reduced demand for skilled handweavers, initially causing reduced wages and unemployment. Protests followed its introduction. For example, in 1816 two thousand rioting Calton weavers tried to destroy power loom mills and stoned the workers. In the longer term, by making cloth more affordable the power loom increased demand and stimulated exports, causing a growth in industrial employment, albeit low-paid. The power loom also opened up opportunities for women mill workers. A darker side of the power loom's impact was the growth of employment of children in power loom mills.

Dangers

When operated by a skilled and attentive weaver, looms are not dangerous by themselves. However, there are a number of inherent dangers in the machines, to which inattentive or poorly trained weavers can fall victim. The most obvious is the moving reed, the frames which hold the heddles and the "pinch" or "sand" roll utilized to keep the cloth tight as it passes over the front of the machine and onto the doff roll. The most common injury in weaving is pinched fingers from distracted or bored workers, though this is not the only such injury found. There are numerous accounts of weavers with long hair getting it tangled in the warp itself and having their scalp pulled away from the skull, or large chunks of hair pulled off. As a result of this, it has become industry standard for companies to require weavers to either keep hair up and tied, or to keep their hair short so as not to allow it to become tangled. Also, due to possible pinch points on the front of machines, loose, baggy clothing is prohibited. One complication for weavers, in the terms of safety, is the loud nature in which weave mills operate. Because of this, it is nearly impossible to hear a person calling for help when entangled, and has led OSHA to outline specific guidelines for companies to mitigate the chance of such circumstances from happening. However, even with such guidelines in place, injuries in textile production, due to the machines themselves, are still commonplace.

Automation

Automation or automatic control, is the use of various control systems for operating equipment such as machinery, processes in factories, boilers and heat treating ovens, switching on tele phone networks, steering and stabilization of ships, aircraft and other applications with minimal or reduced human intervention. Some processes have been completely automated. In modern days, it is also used in our homes. Smart home system is one of the every day application

The biggest benefit of automation is that it saves labor; however, it is also used to save energy and materials and to improve quality, accuracy and precision.

The term *automation*, inspired by the earlier word *automatic* (coming from *automaton*), was not widely used before 1947, when Ford established an automation department. It was during this time that industry was rapidly adopting feedback controllers, which were introduced in the 1930s.

Automation has been achieved by various means including mechanical, hydraulic, pneumatic, electrical, electronic devices and computers, usually in combination. Complicated systems, such as modern factories, airplanes and ships typically use all these combined techniques.

Open-loop and Closed-loop (Feedback) Control

Fundamentally, there are two types of control loop; open loop control, and closed loop (feedback) control.

In open loop control, the control action from the controller is independent of the "process output" (or "controlled process variable"). A good example of this is a central heating boiler controlled only by a timer, so that heat is applied for a constant time, regardless of the temperature of the building. (The control action is the switching on/off of the boiler. The process output is the building temperature).

In closed loop control, the control action from the controller is dependent on the process output. In the case of the boiler analogy this would include a thermostat to monitor the building temperature, and thereby feed back a signal to ensure the controller maintains the building at the temperature set on the thermostat. A closed loop controller therefore has a feedback loop which ensures the controller exerts a control action to give a process output the same as the "Reference input" or "set point". For this reason, closed loop controllers are also called feedback controllers.

The definition of a closed loop control system according to the British Standard Institution is 'a control system possessing monitoring feedback, the deviation signal formed as a result of this feedback being used to control the action of a final control element in such a way as to tend to reduce the deviation to zero.

Likewise; "A *Feedback Control System* is a system which tends to maintain a prescribed relationship of one system variable to another by comparing functions of these variables and using the difference as a means of control.'"

The advanced type of automation that revolutionized manufacturing, aircraft, communications and other industries, is feedback control, which is usually *continuous* and involves taking measurements using a sensor and making calculated adjustments to keep the measured variable within a set range. The theoretical basis of closed loop automation is control theory.

Control Actions

The control action is the form of the controller output action.

Discrete Control (on/off)

One of the simplest types of control is *on-off* control. An example is the thermostat used on household appliances which either opens or closes an electrical contact. (Thermostats were originally developed as true feedback-control mechanisms rather than the on-off common household appliance thermostat.)

Sequence control, in which a programmed sequence of *discrete* operations is performed, often based on system logic that involves system states. An elevator control system is an example of sequence control.

PID Controller

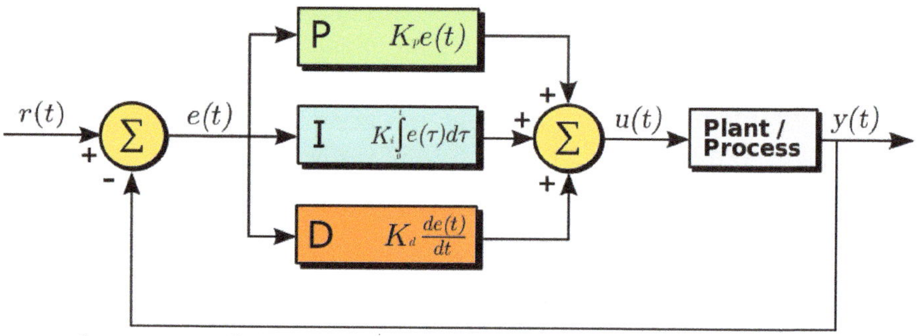

A block diagram of a PID controller in a feedback loop, r(*t*) is the desired process value or "set point", and y(*t*) is the measured process value.

A proportional–integral–derivative controller (PID controller) is a control loop feedback mechanism (controller) widely used in industrial control systems.

A PID controller continuously calculates an *error value* e (t) as the difference between a desired setpoint and a measured process variable and applies a correction based on proportional, integral, and derivative terms, respectively (sometimes denoted *P*, *I*, and *D*) which give their name to the controller type.

The theoretical understanding and application dates from the 1920s, and they are implemented in nearly all analogue control systems; originally in mechanical controllers, and then using discrete electronics and latterly in industrial process computers.

Sequential Control and Logical Sequence or System State Control

Sequential control may be either to a fixed sequence or to a logical one that will perform different actions depending on various system states. An example of an adjustable but otherwise fixed sequence is a timer on a lawn sprinkler.

States refer to the various conditions that can occur in a use or sequence scenario of the system. An example is an elevator, which uses logic based on the system state to perform certain actions in response to its state and operator input. For example, if the operator presses the floor n button, the system will respond depending on whether the elevator is stopped or moving, going up or down, or if the door is open or closed, and other conditions.

An early development of sequential control was relay logic, by which electrical relays engage electrical contacts which either start or interrupt power to a device. Relays were first used in telegraph networks before being developed for controlling other devices, such as when starting and stopping industrial-sized electric motors or opening and closing solenoid valves. Using relays for control purposes allowed event-driven control, where actions could be triggered out of sequence, in response to external events. These were more flexible in their response than the rigid single-sequence cam timers. More complicated examples involved maintaining safe sequences for devices such as swing bridge controls, where a lock bolt needed to be disengaged before the bridge could be moved, and the lock bolt could not be released until the safety gates had already been closed.

State Abstraction

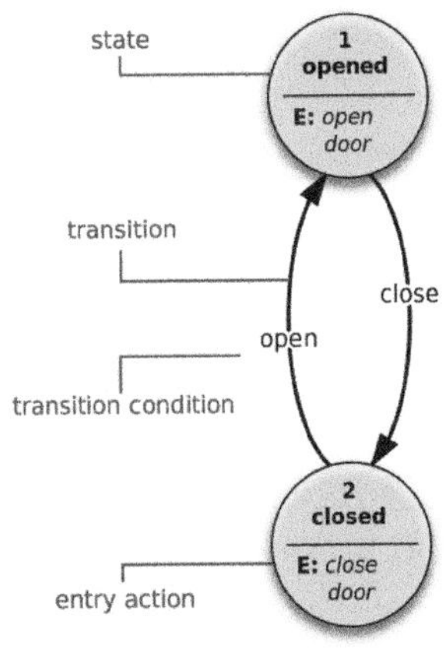

This state diagram shows how UML can be used for designing a door system that can only be opened and closed

The total number of relays, cam timers and drum sequencers can number into the hundreds or even thousands in some factories. Early programming techniques and languages were needed to make such systems manageable, one of the first being ladder logic, where diagrams of the interconnected relays resembled the rungs of a ladder. Special computers called programmable logic controllers were later designed to replace these collections of hardware with a single, more easily re-programmed unit.

In a typical hard wired motor start and stop circuit (called a *control circuit*) a motor is started by pushing a "Start" or "Run" button that activates a pair of electrical relays. The "lock-in" relay locks in contacts that keep the control circuit energized when the push button is released. (The start button is a normally open contact and the stop button is normally closed contact.) Another relay energizes a switch that powers the device that throws the motor starter switch (three sets of contacts for three phase industrial power) in the main power circuit. Large motors use high voltage and experience high in-rush current, making speed important in making and breaking contact. This can be dangerous for personnel and property with manual switches. The "lock in" contacts in the start circuit and the main power contacts for the motor are held engaged by their respective electromagnets until a "stop" or "off" button is pressed, which de-energizes the lock in relay.

Commonly interlocks are added to a control circuit. Suppose that the motor in the example is powering machinery that has a critical need for lubrication. In this case an interlock could be added to insure that the oil pump is running before the motor starts. Timers, limit switches and electric eyes are other common elements in control circuits.

Solenoid valves are widely used on compressed air or hydraulic fluid for powering actuators on mechanical components. While motors are used to supply continuous rotary motion, actuators are typically a better choice for intermittently creating a limited range of movement for a mechanical component, such as moving various mechanical arms, opening or closing valves, raising heavy press rolls, applying pressure to presses.

Computer Control

Computers can perform both sequential control and feedback control, and typically a single computer will do both in an industrial application. Programmable logic controllers (PLCs) are a type of special purpose microprocessor that replaced many hardware components such as timers and drum sequencers used in relay logic type systems. General purpose process control computers have increasingly replaced stand alone controllers, with a single computer able to perform the operations of hundreds of controllers. Process control computers can process data from a network of PLCs, instruments and controllers in order to implement typical (such as PID) control of many individual variables or, in some cases, to implement complex control algorithms using multiple inputs and mathematical manipulations. They can also analyze data and create real time graphical displays for operators and run reports for operators, engineers and management.

Control of an automated teller machine (ATM) is an example of an interactive process in which a computer will perform a logic derived response to a user selection based on information retrieved from a networked database. The ATM process has similarities with other online transaction processes. The different logical responses are called *scenarios*. Such processes are typically designed with the aid of use cases and flowcharts, which guide the writing of the software code.

History

The earliest feedback control mechanism was the thermostat invented in 1620 by the Dutch scientist Cornelius Drebbel. (Note: Early thermostats were temperature regulators or controlers rather than the on-off mechanisms common in household appliances.) Another control mechanism was used to tent the sails of windmills. It was patented by Edmund Lee in 1745. Also in 1745, Jacques de Vaucanson invented the first automated loom.

In 1771 Richard Arkwright invented the first fully automated spinning mill driven by water power, known at the time as the water frame.

The centrifugal governor, which was invented by Christian Huygens in the seventeenth century, was used to adjust the gap between millstones. The centrifugal governor was also used in the automatic flour mill developed by Oliver Evans in 1785, making it the first completely automated industrial process. The governor was adopted by James Watt for use on a steam engine in 1788 after Watt's partner Boulton saw one at a flour mill Boulton & Watt were building.

The governor could not actually hold a set speed; the engine would assume a new constant speed in response to load changes. The governor was able to handle smaller variations such as those caused by fluctuating heat load to the boiler. Also, there was a tendency for oscillation whenever there was a speed change. As a consequence, engines equipped with this governor were not suitable for operations requiring constant speed, such as cotton spinning.

Several improvements to the governor, plus improvements to valve cut-off timing on the steam engine, made the engine suitable for most industrial uses before the end of the 19th century. Advances in the steam engine stayed well ahead of science, both thermodynamics and control theory.

The governor received relatively little scientific attention until James Clerk Maxwell published a paper that established the beginning of a theoretical basis for understanding control theory. Development of the electronic amplifier during the 1920s, which was important for long distance telephony, required a higher signal to noise ratio, which was solved by negative feedback noise cancellation. This and other telephony applications contributed to control theory. Military applications during the Second World War that contributed to and benefited from control theory were fire-control systems and aircraft controls. The word "automation" itself was coined in the 1940s by General Electric. The so-called classical theoretical treatment of control theory dates to the 1940s and 1950s.

Relay logic was introduced with factory electrification, which underwent rapid adaption from 1900 though the 1920s. Central electric power stations were also undergoing rapid growth and operation of new high pressure boilers, steam turbines and electrical substations created a large demand for instruments and controls.

Central control rooms became common in the 1920s, but as late as the early 1930s, most process control was on-off. Operators typically monitored charts drawn by recorders that plotted data from instruments. To make corrections, operators manually opened or closed valves or turned switches on or off. Control rooms also used color coded lights to send signals to workers in the plant to manually make certain changes.

Controllers, which were able to make calculated changes in response to deviations from a set point rather than on-off control, began being introduced the 1930s. Controllers allowed manufacturing to continue showing productivity gains to offset the declining influence of factory electrification.

Factory productivity was greatly increased by electrification in the 1920s. Manufacturing productivity growth fell from 5.2%/yr 1919-29 to 2.76%/yr 1929-41. Field notes that spending on non-medical instruments increased significantly from 1929–33 and remained strong thereafter.

In 1959 Texaco's Port Arthur refinery became the first chemical plant to use digital control. Conversion of factories to digital control began to spread rapidly in the 1970s as the price of computer hardware fell.

Significant Applications

The automatic telephone switchboard was introduced in 1892 along with dial telephones. By 1929, 31.9% of the Bell system was automatic. Automatic telephone switching originally used vacuum tube amplifiers and electro-mechanical switches, which consumed a large amount of electricity. Call volume eventually grew so fast that it was feared the telephone system would consume all electricity production, prompting Bell Labs to begin research on the transistor.

The logic performed by telephone switching relays was the inspiration for the digital computer. The first commercially successful glass bottle blowing machine was an automatic model introduced in 1905. The machine, operated by a two-man crew working 12-hour shifts, could produce 17,280 bottles in 24 hours, compared to 2,880 bottles made by a crew of six men and boys working in a shop for a day. The cost of making bottles by machine was 10 to 12 cents per gross compared to $1.80 per gross by the manual glassblowers and helpers.

Sectional electric drives were developed using control theory. Sectional electric drives are used on different sections of a machine where a precise differential must be maintained between the sections. In steel rolling, the metal elongates as it passes through pairs of rollers, which must run at successively faster speeds. In paper making the paper sheet shrinks as it passes around steam heated drying arranged in groups, which must run at successively slower speeds. The first application of a sectional electric drive was on a paper machine in 1919. One of the most important developments in the steel industry during the 20th century was continuous wide strip rolling, developed by Armco in 1928.

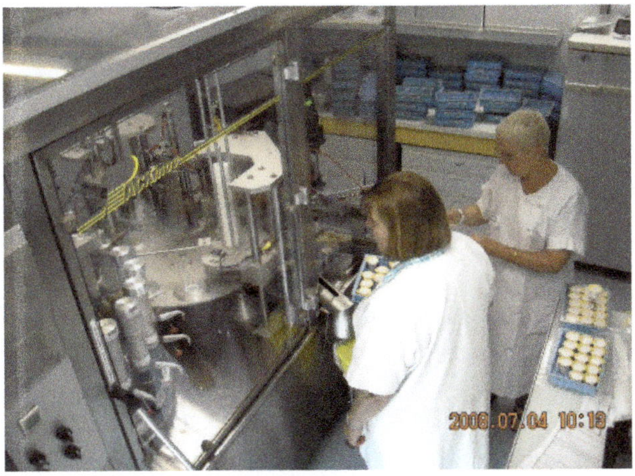

Automated pharmacology production

Before automation many chemicals were made in batches. In 1930, with the widespread use of instruments and the emerging use of controllers, the founder of Dow Chemical Co. was advocating continuous production.

Self-acting machine tools that displaced hand dexterity so they could be operated by boys and unskilled laborers were developed by James Nasmyth in the 1840s. Machine tools were automated with Numerical control (NC) using punched paper tape in the 1950s. This soon evolved into computerized numerical control (CNC).

Today extensive automation is practiced in practically every type of manufacturing and assembly process. Some of the larger processes include electrical power generation, oil refining, chemicals, steel mills, plastics, cement plants, fertilizer plants, pulp and paper mills, automobile and truck assembly, aircraft production, glass manufacturing, natural gas separation plants, food and beverage processing, canning and bottling and manufacture of various kinds of parts. Robots are especially useful in hazardous applications like automobile spray painting. Robots are also used to assemble electronic circuit boards. Automotive welding is done with robots and automatic welders are used in applications like pipelines.

Advantages and Disadvantages

The main advantages of automation are:

- Increased throughput or productivity.
- Improved quality or increased predictability of quality.
- Improved robustness (consistency), of processes or product.
- Increased consistency of output.
- Reduced direct human labor costs and expenses.

The following methods are often employed to improve productivity, quality, or robustness.

- Install automation in operations to reduce cycle time.
- Install automation where a high degree of accuracy is required.
- Replacing human operators in tasks that involve hard physical or monotonous work.
- Replacing humans in tasks done in dangerous environments (i.e. fire, space, volcanoes, nuclear facilities, underwater, etc.)
- Performing tasks that are beyond human capabilities of size, weight, speed, endurance, etc.
- Reduces operation time and work handling time significantly.
- Frees up workers to take on other roles.
- Provides higher level jobs in the development, deployment, maintenance and running of the automated processes.

The main disadvantages of automation are:

- Security Threats/Vulnerability: An automated system may have a limited level of intelligence, and is therefore more susceptible to committing errors outside of its immediate scope of knowledge (e.g., it is typically unable to apply the rules of simple logic to general propositions).
- Unpredictable/excessive development costs: The research and development cost of automating a process may exceed the cost saved by the automation itself.
- High initial cost: The automation of a new product or plant typically requires a very large initial investment in comparison with the unit cost of the product, although the cost of automation may be spread among many products and over time.

In manufacturing, the purpose of automation has shifted to issues broader than productivity, cost, and time.

Lights Out Manufacturing

Lights out manufacturing is when a production system is 100% or near to 100% automated (not hiring any workers). In order to eliminate the need for labor costs altogether.

Health and Environment

The costs of automation to the environment are different depending on the technology, product or engine automated. There are automated engines that consume more energy resources from the Earth in comparison with previous engines and those that do the opposite too. Hazardous operations, such as oil refining, the manufacturing of industrial chemicals, and all forms of metal working, were always early contenders for automation.

Convertibility and Turnaround Time

Another major shift in automation is the increased demand for flexibility and convertibility in manufacturing processes. Manufacturers are increasingly demanding the ability to easily switch from manufacturing Product A to manufacturing Product B without having to completely rebuild the production lines. Flexibility and distributed processes have led to the introduction of Automated Guided Vehicles with Natural Features Navigation.

Digital electronics helped too. Former analogue-based instrumentation was replaced by digital equivalents which can be more accurate and flexible, and offer greater scope for more sophisticated configuration, parametrization and operation. This was accompanied by the fieldbus revolution which provided a networked (i.e. a single cable) means of communicating between control systems and field level instrumentation, eliminating hard-wiring.

Discrete manufacturing plants adopted these technologies fast. The more conservative process industries with their longer plant life cycles have been slower to adopt and analogue-based measurement and control still dominates. The growing use of Industrial Ethernet on the factory floor is pushing these trends still further, enabling manufacturing plants to be integrated more tightly within the enterprise, via the internet if necessary. Global competition has also increased demand for Reconfigurable Manufacturing Systems.

Automation Tools

Engineers can now have numerical control over automated devices. The result has been a rapidly expanding range of applications and human activities. Computer-aided technologies (or CAx) now serve as the basis for mathematical and organizational tools used to create complex systems. Notable examples of CAx include Computer-aided design (CAD software) and Computer-aided manufacturing (CAM software). The improved design, analysis, and manufacture of products enabled by CAx has been beneficial for industry.

Information technology, together with industrial machinery and processes, can assist in the design, implementation, and monitoring of control systems. One example of an industrial control system is a programmable logic controller (PLC). PLCs are specialized hardened computers which are frequently used to synchronize the flow of inputs from (physical) sensors and events with the flow of outputs to actuators and events.

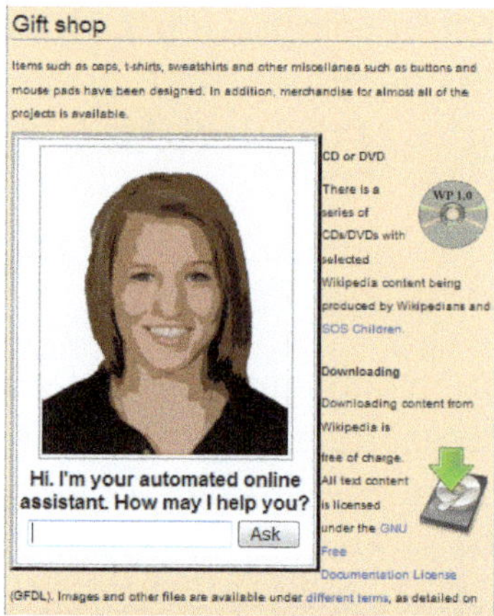

An automated online assistant on a website, with an avatar for enhanced human–computer interaction.

Human-machine interfaces (HMI) or computer human interfaces (CHI), formerly known as *man-machine interfaces*, are usually employed to communicate with PLCs and other computers. Service personnel who monitor and control through HMIs can be called by different names. In industrial process and manufacturing environments, they are called operators or something similar. In boiler houses and central utilities departments they are called stationary engineers.

Different types of automation tools exist:

- ANN - Artificial neural network
- DCS - Distributed Control System
- HMI - Human Machine Interface
- SCADA - Supervisory Control and Data Acquisition
- PLC - Programmable Logic Controller
- Instrumentation
- Motion control
- Robotics

When it comes to Factory Automation, Host Simulation Software (HSS) is a commonly used testing tool that is used to test the equipment software. HSS is used to test equipment performance with respect to Factory Automation standards (timeouts, response time, processing time).

Limitations to Automation

- Current technology is unable to automate all the desired tasks.
- Many operations using automation have large amounts of invested capital and pro-

duce high volumes of product, making malfunctions extremely costly and potentially hazardous. Therefore, some personnel are needed to insure that the entire system functions properly and that safety and product quality are maintained.

- As a process becomes increasingly automated, there is less and less labor to be saved or quality improvement to be gained. This is an example of both diminishing returns and the logistic function.
- As more and more processes become automated, there are fewer remaining non-automated processes. This is an example of exhaustion of opportunities. New technological paradigms may however set new limits that surpass the previous limits.

Current Limitations

Many roles for humans in industrial processes presently lie beyond the scope of automation. Human-level pattern recognition, language comprehension, and language production ability are well beyond the capabilities of modern mechanical and computer systems. Tasks requiring subjective assessment or synthesis of complex sensory data, such as scents and sounds, as well as high-level tasks such as strategic planning, currently require human expertise. In many cases, the use of humans is more cost-effective than mechanical approaches even where automation of industrial tasks is possible. Overcoming these obstacles is a theorized path to post-scarcity economics.

Paradox of Automation

The Paradox of Automation says that the more efficient the automated system, the more crucial the human contribution of the operators. Humans are less involved, but their involvement becomes more critical.

If an automated system has an error, it will multiply that error until it's fixed or shut down. This is where human operators come in.

A fatal example of this was Air France Flight 447, where a failure of automation put the pilots into a manual situation they were not prepared for.

Recent and Emerging Applications

KUKA industrial robots being used at a bakery for food production

Automated Retail

Food and Drink

The food retail industry has started to apply automation to the ordering process; McDonald's has introduced touch screen ordering and payment systems in many of its restaurants, reducing the need for as many cashier employees. The University of Texas at Austin has introduced fully automated cafe retail locations. Some Cafes and restaurants have utilized mobile and tablet "apps" to make the ordering process more efficient by customers ordering and paying on their device. Some restaurants have automated food delivery to customers tables using a Conveyor belt system. The use of robots is sometimes employed to replace waiting staff.

Stores

Many Supermarkets and even smaller stores are rapidly introducing Self checkout systems reducing the need for employing checkout workers.

Online shopping could be considered a form of automated retail as the payment and checkout are through an automated Online transaction processing system. Other forms of automation can also be an integral part of online shopping, for example the deployment of automated warehouse robotics such as that applied by Amazon using Kiva Systems.

Automated Mining

Involves the removal of human labor from the mining process. The mining industry is currently in the transition towards Automation. Currently it can still require a large amount of human capital, particularly in the third world where labor costs are low so there is less incentive for increasing efficiency through automation.

Automated Video Surveillance

The Defense Advanced Research Projects Agency (DARPA) started the research and development of automated visual surveillance and monitoring (VSAM) program, between 1997 and 1999, and airborne video surveillance (AVS) programs, from 1998 to 2002. Currently, there is a major effort underway in the vision community to develop a fully automated tracking surveillance system. Automated video surveillance monitors people and vehicles in real time within a busy environment. Existing automated surveillance systems are based on the environment they are primarily designed to observe, i.e., indoor, outdoor or airborne, the amount of sensors that the automated system can handle and the mobility of sensor, i.e., stationary camera vs. mobile camera. The purpose of a surveillance system is to record properties and trajectories of objects in a given area, generate warnings or notify designated authority in case of occurrence of particular events.

Automated Highway Systems

As demands for safety and mobility have grown and technological possibilities have multiplied, interest in automation has grown. Seeking to accelerate the development and introduction of fully automated vehicles and highways, the United States Congress authorized more than $650 million over six years for intelligent transport systems (ITS) and demonstration projects in the 1991 Intermodal Surface Transportation Efficiency Act (ISTEA). Congress legislated in ISTEA that "the Secretary of Transportation shall develop an automated highway and vehicle prototype from which future fully automated intelligent vehicle-highway systems can be developed. Such development shall include research in human factors to ensure the success of the man-machine relationship. The goal of this program is to have the first fully automated highway roadway or an automated test track in operation by 1997. This system shall accommodate installation of equipment in new and existing motor vehicles."

Full automation commonly defined as requiring no control or very limited control by the driver; such automation would be accomplished through a combination of sensor, computer, and communications systems in vehicles and along the roadway. Fully automated driving would, in theory, allow closer vehicle spacing and higher speeds, which could enhance traffic capacity in places where additional road building is physically impossible, politically unacceptable, or prohibitively expensive. Automated controls also might enhance road safety by reducing the opportunity for driver error, which causes a large share of motor vehicle crashes. Other potential benefits include improved air quality (as a result of more-efficient traffic flows), increased fuel economy, and spin-off technologies generated during research and development related to automated highway systems.

Automated Waste Management

Automated waste collection trucks prevent the need for as many workers as well as easing the level of labor required to provide the service.

Home Automation

Home automation (also called domotics) designates an emerging practice of increased automation of household appliances and features in residential dwellings, particularly through electronic means that allow for things impracticable, overly expensive or simply not possible in recent past decades.

Laboratory Automation

Automation is essential for many scientific and clinical applications. Therefore, automation has been extensively employed in laboratories. From as early as 1980 fully automated laboratories have already been working. However, automation has not become widespread in laboratories due to its high cost. This may change with the ability of integrating low-cost devices with standard laboratory equipment. Autosamplers are common devices used in laboratory automation.

Industrial Automation

Industrial automation deals primarily with the automation of manufacturing, quality control and material handling processes. General purpose controllers for industrial processes include Programmable logic controllers, stand-alone I/O modules, and computers. Industrial automation is to replace the decision making of humans and manual command-response activities with the use of mechanized equipment and logical programming commands. One trend is increased use of Machine vision to provide automatic inspection and robot guidance functions, another is a continuing increase in the use of robots. Industrial automation is simply done at the industrial level.

Energy efficiency in industrial processes has become a higher priority. Semiconductor companies like Infineon Technologies are offering 8-bit micro-controller applications for example found in motor controls, general purpose pumps, fans, and ebikes to reduce energy consumption and thus increase efficiency.

Advantages

- Replaces hard physical or monotonous work
- Tasks in hazardous environments, such as extreme temperatures, or atmospheres that are radioactive or toxic can be done by machines
- Faster production and cheaper labor costs
- Automation can be maintained with simple quality checks.
- Can perform tasks beyond human capabilities.

Disadvantages

- As of now, not all tasks can be automated
- Some tasks are more expensive to automate
- Initial costs are high
- Failure to maintain a system could result in the loss of the product

Industrial Robotics

Industrial robotics is a sub-branch in the industrial automation that aids in various manufacturing processes. Such manufacturing processes include; machining, welding, painting, assembling and material handling to name a few. Industrial robots utilizes various mechanical, electrical as well as software systems to allow for high precision, accuracy and speed that far exceeds any human performance. The birth of industrial robot came shortly after World War II as United States saw the need for a quicker way to produce industrial and consumer goods. Servos, digital logic and solid state electronics allowed engineers to build better and faster systems and overtime these systems were improved and revised to the point where a single robot is capable of running 24 hours a day with little or no maintenance.

Automated milling machines

Programmable Logic Controllers

Industrial automation incorporates programmable logic controllers in the manufacturing process. Programmable logic controllers (PLCs) use a processing system which allows for variation of controls of inputs and outputs using simple programming. PLCs make use of programmable memory, storing instructions and functions like logic, sequencing, timing, counting, etc. Using a logic based language, a PLC can receive a variety of inputs and return a variety of logical outputs, the input devices being sensors and output devices being motors, valves, etc. PLCs are similar to computers, however, while computers are optimized for calculations, PLCs are optimized for control task and use in industrial environments. They are built so that only basic logic-based programming knowledge is needed and to handle vibrations, high temperatures, humidity and noise. The greatest advantage PLCs offer is their flexibility. With the same basic controllers, a PLC can operate a range of different control systems. PLCs make it unnecessary to rewire a system to change the control system. This flexibility leads to a cost-effective system for complex and varied control systems.

Agent-assisted Automation

Agent-assisted automation refers to automation used by call center agents to handle customer inquiries. There are two basic types: desktop automation and automated voice solutions. Desktop automation refers to software programming that makes it easier for the call center agent to work across multiple desktop tools. The automation would take the information entered into one tool and populate it across the others so it did not have to be entered more than once, for example. Automated voice solutions allow the agents to remain on the line while disclosures and other important information is provided to customers in the form of pre-recorded audio files. Specialized applications of these automated voice solutions enable the agents to process credit cards without ever seeing or hearing the credit card numbers or CVV codes

The key benefit of agent-assisted automation is compliance and error-proofing. Agents are sometimes not fully trained or they forget or ignore key steps in the process. The use of automation ensures that what is supposed to happen on the call actually does, every time.

Relationship to Unemployment

Research by the Oxford Martin School showed that employees engaged in "tasks following well-defined procedures that can easily be performed by sophisticated algorithms" are at risk of displacement. The study, published in 2013, shows that automation can affect both skilled and unskilled work and both high and low-paying occupations; however, low-paid physical occupations are most at risk. However, according to a study published in McKinsey Quarterly in 2015 the impact of computerization in most cases is not replacement of employees but automation of portions of the tasks they perform.

Based on a formula by Gilles Saint-Paul, an economist at Toulouse 1 University, the demand for unskilled human capital declines at a slower rate than the demand for skilled human capital increases. In the long run and for society as a whole it has led to cheaper products, lower average work hours, and new industries forming (I.e, robotics industries, computer industries, design industries). These new industries provide many high salary skill based jobs to the economy.

Conveyor System

An overhead chain conveyor conveys cars at Mercedes in Germany

A conveyor system is a common piece of mechanical handling equipment that moves materials from one location to another. Conveyors are especially useful in applications involving the transportation of heavy or bulky materials. Conveyor systems allow quick and efficient transportation for a wide variety of materials, which make them very popular in the material handling and packaging industries. Many kinds of conveying systems are available, and are used according to the various needs of different industries. There are chain conveyors (floor and overhead) as well. Chain conveyors consist of enclosed tracks, I-Beam, towline, power & free, and hand pushed trolleys.

Industries that Use Conveyor Systems

Conveyor systems are used widespread across a range of industries due to the numerous benefits they provide.

- Conveyors are able to safely transport materials from one level to another, which when done by human labor would be strenuous and expensive.

A lineshaft roller conveyor conveys boxed produce at a distribution center

A Conveyor belt conveys papers at a newspaper print plant

- They can be installed almost anywhere, and are much safer than using a forklift or other machine to move materials.

- They can move loads of all shapes, sizes and weights. Also, many have advanced safety features that help prevent accidents.
- There are a variety of options available for running conveying systems, including the hydraulic, mechanical and fully automated systems, which are equipped to fit individual needs.

Roller conveyor for carton transport in the apparel industry

Conveyor systems are commonly used in many industries, including the automotive, agricultural, computer, electronic, food processing, aerospace, pharmaceutical, chemical, bottling and canning, print finishing and packaging. Although a wide variety of materials can be conveyed, some of the most common include food items such as beans and nuts, bottles and cans, automotive components, scrap metal, pills and powders, wood and furniture and grain and animal feed. Many factors are important in the accurate selection of a conveyor system. It is important to know how the conveyor system will be used beforehand. Some individual areas that are helpful to consider are the required conveyor operations, such as transportation, accumulation and sorting, the material sizes, weights and shapes and where the loading and pickup points need to be.

Care and Maintenance of Conveyor Systems

A conveyor system is often the lifeline to a company's ability to effectively move its product in a timely fashion. The steps that a company can take to ensure that it performs at peak capacity, include regular inspections and system audits, close monitoring of motors and reducers, keeping key parts in stock, and proper training of personnel.

Increasing the service life of your conveyor system involves: choosing the right conveyor type, the right system design and paying attention to regular maintenance practices.

A conveyor system that is designed properly will last a long time with proper maintenance. Here are six of the biggest problems to watch for in overhead type conveyor systems including I-beam monorails, enclosed track conveyors and power and free conveyors. Overhead conveyor systems have been used in numerous applications from shop displays, assembly lines to paint finishing plants and more.

Poor take-up adjustment: This is a simple adjustment on most systems yet it is often overlooked. The chain take-up device ensures that the chain is pulled tight as it leaves the drive unit. As wear occurs and the chain lengthens, the take-up extends under the force of its springs. As they extend, the spring force becomes less and the take-up has less effect. Simply compress the take-up springs and your problem goes away. Failure to do this can result in chain surging, jamming, and extreme wear on the track and chain. Take-up adjustment is also important for any conveyor using belts as a means to power rollers, or belts themselves being the mover. With poor-take up on belt-driven rollers, the belt may twist into the drive unit and cause damage, or at the least a noticeable decrease or complete loss of performance may occur. In the case of belt conveyors, a poor take-up may cause drive unit damage or may let the belt slip off of the side of the chassis.

Lack of lubrication: Chain bearings require lubrication in order to reduce friction. The chain pull that the drive experiences can double if the bearings are not lubricated. This can cause the system to overload by either its mechanical or electrical overload protection. On conveyors that go through hot ovens, lubricators can be left on constantly or set to turn on every few cycles.

Contamination: Paint, powder, acid or alkaline fluids, abrasives, glass bead, steel shot, etc. can all lead to rapid deterioration of track and chain. Ask any bearing company about the leading cause of bearing failure and they will point to contamination. Once a foreign substance lands on the raceway of a bearing or on the track, pitting of the surface will occur, and once the surface is compromised, wear will accelerate. Building shrouds around your conveyors can help prevent the ingress of contaminants. Or, pressurize the contained area using a simple fan and duct arrangement. Contamination can also apply to belts (causing slippage, or in the case of some materials premature wear), and of the motors themselves. Since the motors can generate a considerable amount of heat, keeping the surface clean is an almost-free maintenance procedure that can keep heat from getting trapped by dust and grime, which may lead to motor burnout.

Product Handling: In conveyor systems that may be suited for a wide variety of products, such as those in distribution centers, it is important that each new product be deemed acceptable for conveying before being run through the materials handling equipment. Boxes that are too small, too large, too heavy, too light, or too awkwardly shaped may not convey, or may cause many problems including jams, excess wear on conveying equipment, motor overloads, belt breakage, or other damage, and may also consume extra man-hours in terms of picking up cases that slipped between rollers, or damaged product that was not meant for materials handling. If a product such as this manages to make it through most of the system, the sortation system will most likely be the affected, causing jams and failing to properly place items where they are assigned. It should also be noted that any and all cartons handled on any conveyor should be in good shape or spills, jams, downtime, and possible accidents and injuries may result.

Drive Train: Notwithstanding the above, involving take-up adjustment, other parts of the drive train should be kept in proper shape. Broken O-rings on a Lineshaft, pneumatic parts in disrepair, and motor reducers should also be inspected. Loss of power to even one or a few rollers on a conveyor can mean the difference between effective and timely delivery, and repetitive nuances that can continually cost downtime.

Bad Belt Tracking or Timing: In a system that uses precisely controlled belts, such as a sorter system, regular inspections should be made that all belts are traveling at the proper speeds at all times. While usually a computer controls this with Pulse Position Indicators, any belt not controlled must be monitored to ensure accuracy and reduce the likelihood of problems. Timing is also important for any equipment that is instructed to precisely meter out items, such as a merge where one box pulls from all lines at one time. If one were to be mistimed, product would collide and disrupt operation. Timing is also important wherever a conveyor must "keep track" of where a box is, or improper operation will result.

Since a conveyor system is a critical link in a company's ability to move its products in a timely fashion, any disruption of its operation can be costly. Most downtime can be avoided by taking steps to ensure a system operates at peak performance, including regular inspections, close monitoring of motors and reducers, keeping key parts in stock, and proper training of personnel.

Impact & Wear-Resistant Materials Used in Conveyor Systems

Conveyor systems require materials suited to the displacement of heavy loads and the wear-resistance to hold-up over time without seizing due to deformation. In cases, where static control is a factor, special materials designed to either dissipate or conduct electrical charges. Example of conveyor handling materials include UHMW, Nylon, Nylatron NSM, HDPE, Tivar, Tivar ESd, and Polyurethane

Growth of Conveyor Systems in Various Industries

As far as growth is concerned the material handling and conveyor system makers are getting utmost exposure in the industries like automotive, pharmaceutical, packaging and different production plants. The portable conveyors are likewise growing fast in the construction sector and by the year 2014 the purchase rate for conveyor systems in North America, Europe and Asia is likely to grow even further. Mostly purchased conveyor equipments are Line shaft roller conveyor, chain conveyors and conveyor belts at packaging factories and industrial plants where usually product finishing and monitoring are carried. Commercial and civil sectors are increasingly implementing conveyors at airports, shopping malls, etc. The increasing construction of malls and airports around world shows positive scope and growth for manufacturers of conveyor belts

Types of Conveyor Systems

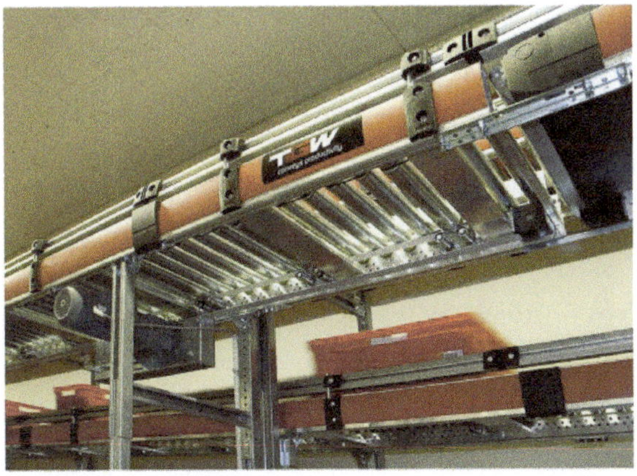

Belt driven roller conveyor for cartons and totes.

Flexible conveyor

- Gravity conveyor
- Gravity skatewheel conveyor
- Belt conveyor
- Wire mesh conveyors
- Plastic belt conveyors
- Bucket conveyors
- Flexible conveyors
- Vertical conveyors
- Spiral conveyors
- Vibrating conveyors

- Pneumatic conveyors
- Electric track vehicle systems
- Belt driven live roller conveyors
- Lineshaft roller conveyor
- Chain conveyor
- Screw conveyor or auger conveyor
- Chain driven live roller conveyor
- Overhead I-beam conveyors
- Dust proof conveyors
- Pharmaceutical conveyors
- Automotive conveyors
- Overland conveyor
- Drag Conveyor

Pneumatic Conveyor Systems

Every pneumatic system makes use of pipes or ducts called transportation lines that carry a mixture of materials and a stream of air. These materials are such as dry pulverized or free flowing or light powdery materials like cement, fly ash etc. These materials can be transported conveniently to various destinations by means of a stream of high velocity air through pipe lines. Products are moved through various tubes via air pressure, allowing for extra vertical versatility. Pneumatic conveyors are either carrier systems or dilute-phase systems; carrier systems simply push items from one entry point to one exit point, such as the money-exchanging pneumatic tubes used at a bank drive-through window. Dilute-phase systems use push-pull pressure to guide materials through various entry and/or exit points. It is important to note that either air compressors, vacuums, or blowers can be used to generate the air flow. This will all depend on what the engineers think will be the most efficient and economical way of developing the system. Three basic systems that are used to generate high velocity air stream:

1. Suction or vacuum systems, utilizing a vacuum created in the pipeline to draw the material with the surrounding air.The system operated at a low pressure, which is practically 0.4–0.5 atm below atmosphere, and is utilized mainly in conveying light free flowing materials.
2. Pressure-type systems, in which a positive pressure is used to push material

from one point to the next. The system is ideal for conveying material from one loading point to a number of unloading points. It operates at a pressure of 6 atm and upwards.

3. Combination systems, in which a suction system is used to convey material from a number of loading points and a pressure system is employed to deliver it to a number of unloading points.

Vibrating Conveyor Systems

A Vibrating Conveyor is a machine with a solid conveying surface which is turned up on the side to form a trough. They are used extensively in food grade applications where sanitation, washdown, and low maintenance are essential. Vibrating conveyors are also suitable for harsh, very hot, dirty, or corrosive environments. They can be used to convey newly cast metal parts which may reach upwards of 1,500 °F (820 °C). Due to the fixed nature of the conveying pans vibrating conveyors can also perform tasks such as sorting, screening, classifying and orienting parts. Vibrating conveyors have been built to convey material at angles exceeding 45° from horizontal using special pan shapes. Flat pans will convey most materials at a 5° Incline from horizontal line.

Flexible Conveyor Systems

The flexible conveyor is based on a conveyor beam in aluminium or stainless steel, with low friction slide rails guiding a plastic multi-flexing chain. Products to be conveyed travel directly on the conveyor, or on pallets/carriers. These conveyors can be worked around obstacles and keep production lines flowing. They are made at varying levels and can work in multiple environments. They are used in food packaging, case packing, and pharmaceutical industries but also in retail stores such as Wal-Mart and Kmart.

Vertical Conveyor Systems and Spiral Conveyors

Vertical conveyor - also commonly referred to as freight lifts and material lifts - are conveyor systems used to raise or lower materials to different levels of a facility during the handling process. Examples of these conveyors applied in the industrial assembly process include transporting materials to different floors. While similar in look to freight elevators, vertical conveyors are not equipped to transport people, only materials.

Vertical lift conveyors contain two adjacent, parallel conveyors for simultaneous upward movement of adjacent surfaces of the parallel conveyors. One of the conveyors normally has spaced apart flites for transporting bulk food items. The dual conveyors rotate in opposite directions, but are operated from one gear box to insure equal belt speed. One of the conveyors is pivotally hinged to the other conveyor for swinging the pivotally attached conveyor away from the remaining conveyor for access to the facing surfaces of the parallel conveyors. Vertical lift conveyors can be manually or automatically loaded and controlled. Almost all vertical conveyors can be systematically integrated with horizontal conveyors, since both of these conveyor systems work in tandem to create a cohesive material handling assembly line.

In similarity to vertical conveyors, spiral conveyors raise and lower materials to different levels of a facility. In contrast, spiral conveyors are able to transport material loads in a continuous flow. Industries that require a higher output of materials - food and beverage, retail case packaging, pharmaceuticals - typically incorporate these conveyors into their systems over standard vertical conveyors due to their ability to facilitate high throughput. Most spiral conveyors also have a lower angle of incline or decline (11 degrees or less) to prevent sliding and tumbling during operation.

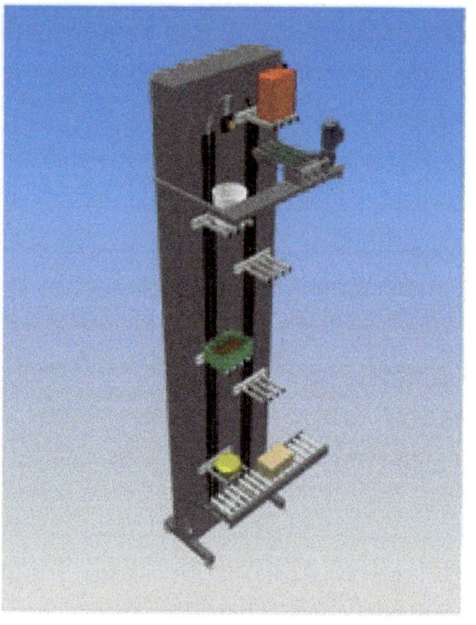

Vertical conveyor with forks

Just like spiral conveyors also a vertical conveyor that use forks are able to transport material loads in a continuous flow. With these forks the load can be taken from one horizontal conveyor and put down on another horizontal conveyor on a different level. By adding more forks more products can be lifted at the same time. Conventional vertical conveyors have the restriction that the input and output of material loads must have the same direction. By using forks many combinations of different input- and outputlevels in different directions are possible. A vertical conveyor with forks can even be used as a vertical sorter. Compared to a spiral conveyor a vertical conveyor - with or without forks - takes up less space.

Heavy Duty Rear Conveyors

Heavy Duty roller conveyors are used for moving items that are at least 500 lbs. This type of conveyor makes the handling of such heavy equipment/products easier and more time effective. Many of the heavy duty roller conveyors can move as fast as 75 feet/minute.

Other types of heavy duty roller conveyors are gravity roller conveyor, chain driven live roller conveyor, pallet accumulation conveyor, multi-strand chain conveyor, and chain & roller transfers.

Gravity roller conveyors are extremely easy to use and are used in many different types of industries such as automotive and retail.

Chain driven live roller conveyors are used for single or bi-directional material handling. Large heavy loads are moved by chain driven live roller conveyors.

Pallet accumulation conveyors are powered through a mechanical clutch. This is used instead of individually powered and controlled sections of conveyors.

Multi-strand chain conveyors are used for double pitch roller chains. Products that can not be moved on traditional roller conveyors can be moved by a multi-strand chain conveyor.

Chain & roller conveyors are short runs of two or more strands of double pitch chain conveyor built into a chain driven line roller conveyor. These pop up under the load and move the load off of the conveyor.

Conveyor Belt

Point of contact between a power transmission belt and its pulley. A conveyor belt uses a wide belt and pulleys and is supported by rollers or a flat pan along its path.

A conveyor belt is the carrying medium of a belt conveyor system (often shortened to belt conveyor). A belt conveyor system is one of many types of conveyor systems. A belt conveyor system consists of two or more pulleys (sometimes referred to as drums), with an endless loop of carrying medium—the conveyor belt—that rotates about them. One or both of the pulleys are powered, moving the belt and the material on the belt forward. The powered pulley is called the drive pulley while the unpowered pulley is called the idler pulley. There are two main industrial classes of belt conveyors; Those in general material handling such as those moving boxes along inside a factory and bulk material handling such as those used to transport large volumes of resources and agricultural materials, such as grain, salt, coal, ore, sand, overburden and more.

Today there are different types of conveyor belts that have been created for conveying different kinds of material available in PVC and rubber materials.

The belt consists of one or more layers of material. Many belts in general material handling have two layers. An under layer of material to provide linear strength and shape called a carcass and an over layer called the cover. The carcass is often a woven fabric having a warp & weft. The most common carcass materials are polyester, nylon and cotton. The cover is often various rubber or plastic compounds specified by use of the belt. Covers can be made from more exotic materials for unusual applications such as silicone for heat or gum rubber when traction is essential.

These conveyor structures contain belts for moving bulk sulfur from railcars to storage piles and from the piles to ships.

A conveyor belt can be a slide and be controlled by the force of gravity.

Material flowing over the belt may be weighed in transit using a beltweigher. Belts with regularly spaced partitions, known as *elevator belts*, are used for transporting loose materials up steep inclines. Belt Conveyors are used in self-unloading bulk freighters and in live bottom trucks. Belt conveyor technology is also used in conveyor transport such as moving sidewalks or escalators, as well as on many manufacturing assembly lines. Stores often have conveyor belts at the check-out counter to move shopping items. Ski areas also use conveyor belts to transport skiers up the hill.

Some of the major global conveyor belt service providers are Fenner plc, Terra Nova Technologies, ThyssenKrupp, HESE Maschinenfabrik GmbH and Tenova Takraf.

Belt Conveyor Systems

Conveyors are durable and reliable components used in automated distribution and warehousing. In combination with computer controlled pallet handling equipment this allows for more efficient retail, wholesale, and manufacturing distribution. It is considered a labor saving system that allows large volumes to move rapidly through a process, allowing companies to ship or receive higher volumes with smaller storage space and with less labor expense.

Rubber conveyor belts are commonly used to convey items with irregular bottom surfaces, small items that would fall in between rollers (e.g. a sushi conveyor bar), or bags of product that would sag between rollers. Belt conveyors are generally fairly similar in construction consisting of a metal frame with rollers at either end of a flat metal bed. The belt is looped around each of the rollers and when one of the rollers is powered (by an electrical motor) the belting slides across the solid metal frame bed, moving the product. In heavy use applications the beds which the belting is pulled over are replaced with rollers. The rollers allow weight to be conveyed as they reduce the amount of friction generated from the heavier loading on the belting.

Belt conveyors can now be manufactured with curved sections which use tapered rollers and curved belting to convey products around a corner. These conveyor systems are commonly used in postal sorting offices and airport baggage handling systems. A sandwich belt conveyor uses two conveyor belts, face-to-face, to firmly contain the item being carried, making steep incline and even vertical-lift runs achievable.

Belt conveyors are the most commonly used powered conveyors because they are the most versatile and the least expensive. Product is conveyed directly on the belt so both regular and irregular shaped objects, large or small, light and heavy, can be transported successfully. These conveyors should use only the highest quality premium belting products, which reduces belt stretch and results in less maintenance for tension adjustments. Belt conveyors can be used to transport product in a straight line or through changes in elevation or direction. In certain applications they can also be used for static accumulation or cartons.

Belt Conveyor systems at a Packing Depot

Baggage Handling Belt Conveyor systems

Belt conveyor from an open-pit lignite mine in Bavaria

Long Belt Conveyors

Taranto, Paolo Monti, 1964

The longest belt conveyor system in the world is in Western Sahara. It is 98 km (61 mi) long, from the phosphate mines of Bu Craa to the coast south of El-Aaiun.

The longest conveyor system in an airport is the Dubai International Airport baggage handling system at 63 km (39 mi). It was installed by Siemens and commissioned in 2008, and has a combination of traditional belt conveyors and tray conveyors.

Boddington Bauxite Mine in Western Australia is officially recognised as having the world's longest and second-longest single belts with a 31-kilometre-long (19 mi) belt feeding a 20 km (12.5 miles) long belt. This system feeds bauxite through the difficult terrain of the Darling Ranges to the Worsley Alumina refinery. The longest single-belt international conveyor runs from Meghalaya in India to a cement factory at Chhatak Bangladesh. It is about 17 km long and conveys limestone and shale at 960 tons/hour, from the quarry in India to the cement factory (7 km long in India and 10 km long in Bangladesh). The conveyor was engineered by AUMUND France and Larsen & Toubro. The conveyor is actuated by three synchronized drive units for a total power of about 1.8 MW supplied by ABB (two drives at the head end in Bangladesh and one drive at the tail end in India). The conveyor belt was manufactured in 300-meter lengths on the Indian side and 500-meter lengths on the Bangladesh side, and was installed on-site by NILOS India. The idlers, or rollers, of the system are unique in that they are designed to accommodate both horizontal and vertical curves along the terrain. Dedicated vehicles were designed for the maintenance of the conveyor, which is always at a minimum height of 5 metres (16 ft) above the ground to avoid being flooded during monsoon periods.

Belt Conveyor Safety System

Conveyors used in industrial settings include tripping mechanisms such as trip cords along the length of the conveyor. This allows for workers to immediately shut down the conveyor when a problem arises. Warning alarms are included to notify employees that a conveyor is about to turn on. In the United States, the Occupational Safety and Health Administration has issued regulations for conveyor safety, as OSHA 1926.555.

History

Primitive conveyor belts were used since the 19th century. In 1892, Thomas Robins began a series of inventions which led to the development of a conveyor belt used for carrying coal, ores and other products. In 1901, Sandvik invented and started the production of steel conveyor belts. In 1905 Richard Sutcliffe invented the first conveyor belts for use in coal mines which revolutionized the mining industry. In 1913, Henry Ford introduced conveyor-belt assembly lines at Ford Motor Company's Highland Park, Michigan factory. In 1972, the French society REI created in New Caledonia the then longest straight-belt conveyor in the world, at a length of 13.8 km. Hyacynthe Marcel Bocchetti was the concept designer. In 1957, the B. F. Goodrich Company patented a conveyor belt that it went on to produce as the Turnover Conveyor Belt System. Incorporating a half-twist, it had the advantage over conventional belts of a longer life because it could expose all of its surface area to wear and tear. Möbius strip belts are no longer manufactured because untwisted modern belts can be made more durable by constructing them from several layers of different materials. In 1970, Intralox, a Louisiana-based company, registered the first patent for all plastic, modular belting.

Flexible Manufacturing System

A typical FMS.

A flexible manufacturing system (FMS) is a manufacturing system in which there is some amount of flexibility that allows the system to react in case of changes, whether predicted or unpredicted. This flexibility is generally considered to fall into two categories, which both contain numerous subcategories.

The first category, *machine flexibility*, covers the system's ability to be changed to produce new product types, and ability to change the order of operations executed on a part. The second category is called *routing flexibility*, which consists of the ability to use multiple machines to perform the same operation on a part, as well as the system's ability to absorb large-scale changes, such as in volume, capacity, or capability.

Most FMS consist of three main systems. The work machines which are often automated CNC machines are connected by a material handling system to optimize parts flow and the central control computer which controls material movements and machine flow.

The main advantages of an FMS is its high flexibility in managing manufacturing resources like time and effort in order to manufacture a new product. The best application of an FMS is found in the production of small sets of products like those from a mass production.

Advantages

- Reduced manufacturing cost
- Lower cost per unit produced,
- Greater labour productivity,
- Greater machine efficiency,

- Improved quality,
- Increased system reliability,
- Reduced parts inventories,
- Adaptability to CAD/CAM operations.
- Shorter lead times
- Improved efficiency
- Increase production rate

Disadvantages

- Initial set-up cost is high,
- Substantial pre-planning
- Requirement of skilled labour
- Complicated system

Flexibility

Flexibility in manufacturing means the ability to deal with slightly or greatly mixed parts, to allow variation in parts assembly and variations in process sequence, change the production volume and change the design of certain product being manufactured.

Industrial FMS Communication

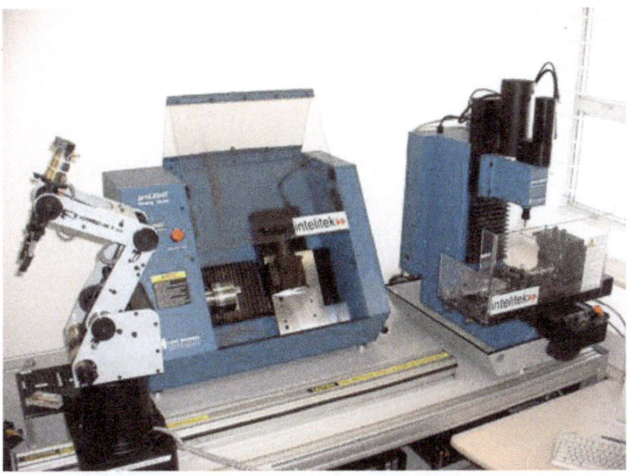

Training FMS with learning robot SCORBOT-ER 4u, workbench CNC Mill and CNC Lathe

An Industrial Flexible Manufacturing System (FMS) consists of robots, Computer-controlled Machines, Numerical controlled machines (CNC), instrumentation devices, computers, sensors, and other stand alone systems such as inspection machines. The use of robots in the production segment of manufacturing industries promises a variety of benefits ranging from high utilization to high volume of productivity. Each Robotic cell or node will be located along a material handling system such as a conveyor or automatic guided vehicle. The production of each part or work-piece will require a different combination of manufacturing nodes. The movement of parts from one node to another is done through the material handling system. At the end of part processing, the finished parts will be routed to an automatic inspection node, and subsequently unloaded from the Flexible Manufacturing System.

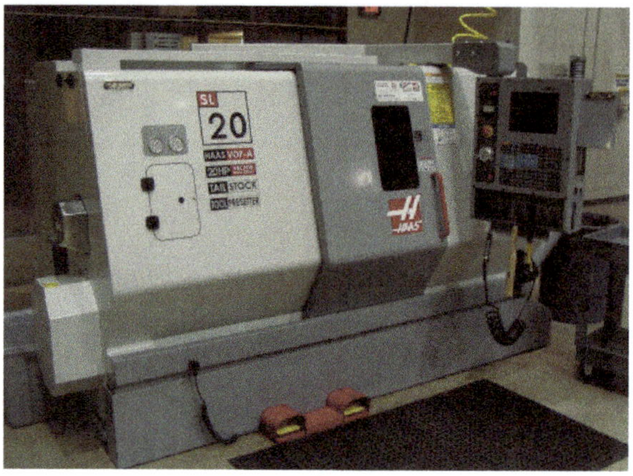

CNC machine

The FMS data traffic consists of large files and short messages, and mostly come from nodes, devices and instruments. The message size ranges between a few bytes to several hundreds of bytes. Executive software and other data, for example, are files with a large size, while messages for machining data, instrument to instrument communications, status monitoring, and data reporting are transmitted in small size.

There is also some variation on response time. Large program files from a main computer usually take about 60 seconds to be down loaded into each instrument or node at the beginning of FMS operation. Messages for instrument data need to be sent in a periodic time with deterministic time delay. Other types of messages used for emergency reporting are quite short in size and must be transmitted and received with an almost instantaneous response. The demands for reliable FMS protocol that support all the FMS data characteristics are now urgent. The existing IEEE standard protocols do not fully satisfy the real time communication requirements in this environment. The delay of CSMA/CD is unbounded as the number of nodes increases due to the message collisions. Token Bus has a deterministic message delay, but it does not support prioritized access scheme which is needed in FMS communications. Token Ring provides prioritized access and has a low message delay, however, its data transmission is unreliable. A single node failure which may occur quite often in FMS causes transmission errors of passing message in that node. In addition, the topology of Token Ring results in high wiring installation and cost.

A design of FMS communication that supports a real time communication with bounded message delay and reacts promptly to any emergency signal is needed. Because of machine failure and malfunction due to heat, dust, and electromagnetic interference is common, a prioritized mechanism and immediate transmission of emergency messages are needed so that a suitable recovery procedure can be applied. A modification of standard Token Bus to implement a prioritized access scheme was proposed to allow transmission of short and periodic messages with a low delay compared to the one for long messages.

Computer-Aided Manufacturing

Computer-aided manufacturing (CAM) is the use of software to control machine tools and related ones in the manufacturing of workpieces. This is not the only definition for CAM, but it is the most common; CAM may also refer to the use of a computer to assist in all operations of a manufacturing plant, including planning, management, transportation and storage. Its primary purpose is to create a faster production process and components and tooling with more precise dimensions and material consistency, which in some cases, uses only the required amount of raw material (thus minimizing waste), while simultaneously reducing energy consumption. CAM is now a system used in schools and lower educational purposes. CAM is a subsequent computer-aided process after computer-aided design (CAD) and sometimes computer-aided engineering (CAE), as the model generated in CAD and verified in CAE can be input into CAM software, which then controls the machine tool. CAM is used in many schools alongside computer-aided design (CAD) to create objects.

Overview

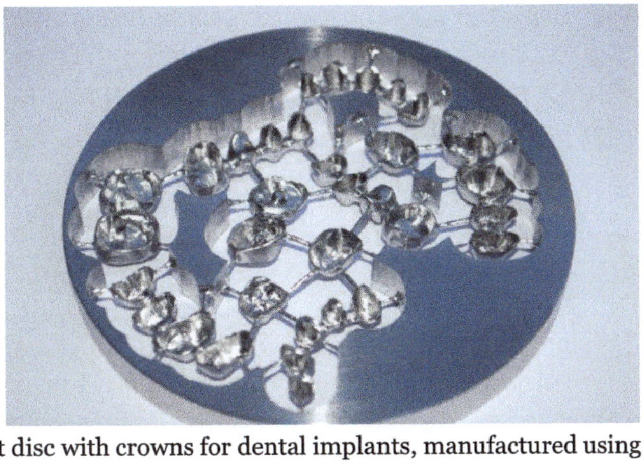

Chrome-cobalt disc with crowns for dental implants, manufactured using WorkNC CAM

Traditionally, CAM has been considered as a numerical control (NC) programming tool, where in two-dimensional (2-D) or three-dimensional (3-D) models of components generated in CADAs with other "Computer-Aided" technologies, CAM does not eliminate the need for skilled professionals such as manufacturing engineers, NC programmers, or machinists. CAM, in fact, leverages both the value of the most skilled manufacturing professionals through advanced productivity tools, while building the skills of new professionals through visualization, simulation and optimization tools.

History

Early commercial applications of CAM was in large companies in the automotive and aerospace industries, for example Pierre Béziers work developing the CAD/CAM application UNISURF in the 1960s for car body design and tooling at Renault.

Historically, CAM software was seen to have several shortcomings that necessitated an overly high level of involvement by skilled CNC machinists. Fallows created the first CAD software but this had severe shortcomings and was promptly taken back into the developing stage. CAM software would output code for the least capable machine, as each machine tool control added on to the standard G-code set for increased flexibility. In some cases, such as improperly set up CAM software or specific tools, the CNC machine required manual editing before the program will run properly. None of these issues were so insurmountable that a thoughtful engineer or skilled machine operator could not overcome for prototyping or small production runs; G-Code is a simple language. In high production or high precision shops, a different set of problems were encountered where an experienced CNC machinist must both hand-code programs and run CAM software.

Integration of CAD with other components of CAD/CAM/CAE Product lifecycle management (PLM) environment requires an effective CAD data exchange. Usually it had been necessary to force the CAD operator to export the data in one of the common data formats, such as IGES or STL or Parasolid formats that are supported by a wide variety of software. The output from the CAM software is usually a simple text file of G-code/ M-codes, sometimes many thousands of commands long, that is then transferred to a machine tool using a direct numerical control (DNC) program or in modern Controllers using a common USB Storage Device.

CAM packages could not, and still cannot, reason as a machinist can. They could not optimize toolpaths to the extent required of mass production. Users would select the type of tool, machining process and paths to be used. While an engineer may have a working knowledge of G-code programming, small optimization and wear issues compound over time. Mass-produced items that require machining are often initially created through casting or some other non-machine method. This enables hand-written, short, and highly optimized G-code that could not be produced in a CAM package.

At least in the United States, there is a shortage of young, skilled machinists entering the workforce able to perform at the extremes of manufacturing; high precision and mass production. As CAM software and machines become more complicated, the skills required of a machinist or machine operator advance to approach that of a computer programmer and engineer rather than eliminating the CNC machinist from the workforce.

Typical Areas of Concern:

- High Speed Machining, including streamlining of tool paths
- Multi-function Machining
- 5 Axis Machining
- Feature recognition and machining
- Automation of Machining processes
- Ease of Use

Overcoming Historical Shortcomings

Over time, the historical shortcomings of CAM are being attenuated, both by providers of niche solutions and by providers of high-end solutions. This is occurring primarily in three arenas:

1. Ease of usage
2. Manufacturing complexity
3. Integration with PLM and the extended enterprise

Ease in use

For the user who is just getting started as a CAM user, out-of-the-box capabilities providing Process Wizards, templates, libraries, machine tool kits, automated feature based machining and job function specific tailorable user interfaces build user confidence and speed the learning curve.

User confidence is further built on 3D visualization through a closer integration with the 3D CAD environment, including error-avoiding simulations and optimizations.

Manufacturing complexity

The manufacturing environment is increasingly complex. The need for CAM and PLM tools by buMs are NC programmer or machinist is similar to the need for computer assistance by the pilot of modern aircraft systems. The modern machinery cannot be properly used without this assistance.

Today's CAM systems support the full range of machine tools including: turning, 5 axis machining and wire EDM. Today's CAM user can easily generate streamlined tool paths, optimized tool axis tilt for higher feed rates, better tool life and surface finish and optimized Z axis depth cuts as well as driving non-cutting operations such as the specification of probing motions.

Integration with PLM and the extended enterpriseLM to integrate manufacturing with enterprise operations from concept through field support of the finished product.

To ensure ease of use appropriate to user objectives, modern CAM solutions are scalable from a stand-alone CAM system to a fully integrated multi-CAD 3D solution-set. These solutions are created to meet the full needs of manufacturing personnel including part planning, shop documentation, resource management and data management and exchange. To prevent these solutions from detailed tool specific information a dedicated tool management

Machining Process

Most machining progresses through many stages, each of which is implemented by a variety of basic and sophisticated strategies, depending on the material and the software available.

Roughing

This process begins with raw stock, known as billet, and cuts it very roughly to shape of the final model. In milling, the result often gives the appearance of terraces, because the strategy has taken advantage of the ability to cut the model horizontally. Common strategies are zig-zag clearing, offset clearing, plunge roughing, rest-roughing.

Semi-f

This process begins with a roughed part that unevenly approximates the model and cuts to within a fixed offset distance from the model. The semi-finishing pass must leave a small amount of material so the tool can cut accurately while finishing, but not so little that the tool and material deflect instead of sending. Common strategies are raster passes, waterline passes, constant step-over passes, pencil milling.

Finishing

Finishing involves a slow pass across the material in very fine steps to produce the finished part. In finishing, the step between one pass and another is minimal. Feed rates are low and spindle speeds are raised to produce an accurate surface.

Contour milling

In milling applications on hardware with five or more axes, a separate finishing process called contouring can be performed. Instead of stepping down in fine-grained increments to approximate a surface, the work piece is rotated to make the cutting surfaces of the tool tangent to the ideal part features. This produces an excellent surface finish with high dimensional accuracy.

Software: Large Vendors

The top 20 largest CAM software companies, by direct revenues in year 2015, are sorted by global revenues:

- - CATIA, Dassault Systèmes
- - Solid Edge, Siemens
- - edgecam, work NC, surfcam Vero Software
- - HSM, Powermill, Featurecam, AUTODESK
- - CAMWorks, Geometric Technologies
- - HyperMill, OPEN MIND Technologies
- - Tebis
- - Mastercam, CNC Software
- - Cimatron, GibbsCAM 3D Systems
- - Creo, PTC
- - Vericut, CG Tech
- - TopSolid, Missler Software
- - Sprut Technology
- - FlexiSign, SAI Software
- - TYPE3, Gravotech Group
- - MecSoft Corporation
- - C&G Systems
- - SolidCAM, SolidCAM
- - NTT Data Engineering Systems
- - BobCAD-CAM

Computer-Integrated Manufacturing

Computer-integrated manufacturing (CIM) is the manufacturing approach of using computers to control the entire production process. This integration allows individual processes to exchange information with each other and initiate actions. Although manufacturing can be faster and less error-prone by the integration of computers, the main advantage is the ability to create automated manufacturing processes. Typically CIM relies on closed-loop control processes, based on real-time input from sensors. It is also known as *flexible design and manufacturing*.

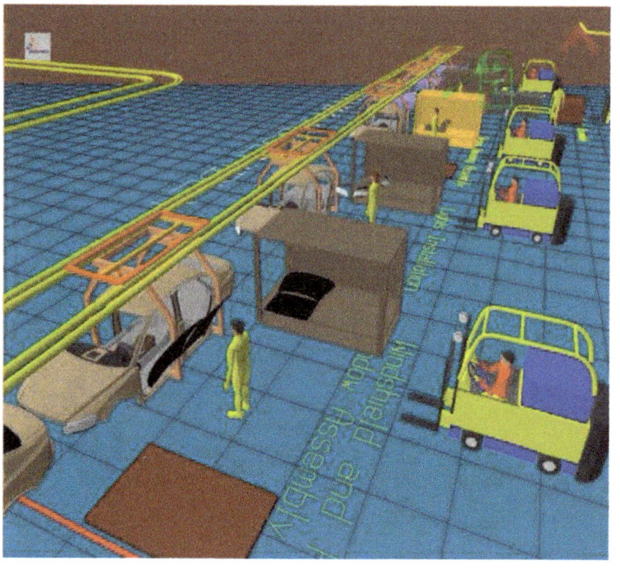

Manufacturing Systems Integration Program, NIST 2008.

Overview

Computer-integrated manufacturing is used in automotive, aviation, space, and ship building industries. The term "computer-integrated manufacturing" is both a method of manufacturing and the name of a computer-automated system in which individual engineering, production, marketing, and support functions of a manufacturing enterprise are organized. In a CIM system functional areas such as design, analysis, planning, purchasing, cost accounting, inventory control, and distribution are linked through the computer with factory floor functions such as materials handling and management, providing direct control and monitoring of all the operations.

As a method of manufacturing, three components distinguish CIM from other manufacturing methodologies:

- Means for data storage, retrieval, manipulation and presentation;
- Mechanisms for sensing state and modifying processes;

- Algorithms for uniting the data processing component with the sensor/modification component.

CIM is an example of the implementation of information and communication technologies (ICTs) in manufacturing.

CIM implies that there are at least two computers exchanging information, e.g. the controller of an arm robot and a micro-controller of a CNC machine.

Some factors involved when considering a CIM implementation are the production volume, the experience of the company or personnel to make the integration, the level of the integration into the product itself and the integration of the production processes. CIM is most useful where a high level of ICT is used in the company or facility, such as CAD/CAM systems, the availability of process planning and its data.

History

The idea of "digital manufacturing" became prominent in the early 1970s, with the release of Dr. Joseph Harrington's book, Computer Integrated Manufacturing. However,it was not until 1984 when computer-integrated manufacturing began to be developed and promoted by machine tool manufacturers and the Computer and Automated Systems Association and Society of Manufacturing Engineers (CASA/SME).

> *"CIM is the integration of total manufacturing enterprise by using integrated systems and data communication coupled with new managerial philosophies that improve organizational and personnel efficiency."* ERHUM

Topics

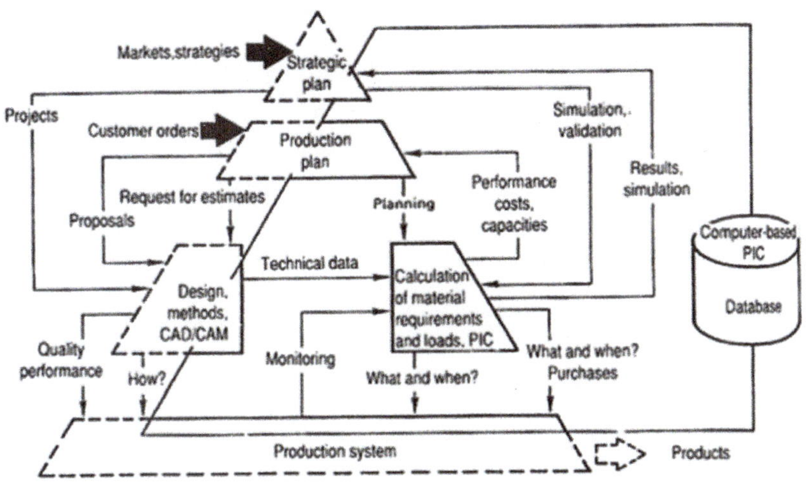

CIM & production control system: Computer Integrated Manufacturing is used to describe the complete automation of a manufacturing plant, with all processes running under computer control and digital information tying them together.

Key Challenges

There are three major challenges to development of a smoothly operating computer-integrated manufacturing system:

- Integration of components from different suppliers: When different machines, such as CNC, conveyors and robots, are using different communications protocols (In the case of AGVs, even differing lengths of time for charging the batteries) may cause problems.
- Data integrity: The higher the degree of automation, the more critical is the integrity of the data used to control the machines. While the CIM system saves on labor of operating the machines, it requires extra human labor in ensuring that there are proper safeguards for the data signals that are used to control the machines.
- Process control: Computers may be used to *assist* the human operators of the manufacturing facility, but there must always be a competent engineer on hand to handle circumstances which could not be foreseen by the designers of the control software.

Subsystems

A computer-integrated manufacturing system is not the same as a *"lights-out factory"*, which would run completely independent of human intervention, although it is a big step in that direction. Part of the system involves flexible manufacturing, where the factory can be quickly modified to produce different products, or where the volume of products can be changed quickly with the aid of computers. Some or all of the following subsystems may be found in a CIM operation:

Computer-aided techniques:

- CAD (computer-aided design)
- CAE (computer-aided engineering)
- CAM (computer-aided manufacturing)
- CAPP (computer-aided process planning)
- CAQ (computer-aided quality assurance)
- PPC (production planning and control)
- ERP (enterprise resource planning)
- A business system integrated by a common database.

Devices and equipment required:

- CNC, Computer numerical controlled machine tools
- DNC, Direct numerical control machine tools
- PLCs, Programmable logic controllers
- Robotics
- Computers
- Software
- Controllers
- Networks
- Interfacing
- Monitoring equipment

Technologies:

- FMS, (flexible manufacturing system)
- ASRS, automated storage and retrieval system
- AGV, automated guided vehicle
- Robotics
- Automated conveyance systems

Others:

- Lean manufacturing

CIMOSA

CIMOSA (Computer Integrated Manufacturing Open System Architecture), is a 1990s European proposal for an open systems architecture for CIM developed by the AMICE Consortium as a series of ESPRIT projects. The goal of CIMOSA was "to help companies to manage change and integrate their facilities and operations to face world wide competition. It provides a consistent architectural framework for both enterprise modeling and enterprise integration as required in CIM environments".

CIMOSA provides a solution for business integration with four types of products:

- The CIMOSA Enterprise Modeling Framework, which provides a reference architecture for enterprise architecture
- CIMOSA IIS, a standard for physical and application integration.

- CIMOSA Systems Life Cycle, is a life cycle model for CIM development and deployment.
- Inputs to standardization, basics for international standard development.

CIMOSA according to Vernadat (1996), coined the term business process and introduced the process-based approach for integrated enterprise modeling based on a cross-boundaries approach, which opposed to traditional function or activity-based approaches. With CIMOSA also the concept of an “Open System Architecture” (OSA) for CIM was introduced, which was designed to be vendor-independent, and constructed with standardised CIM modules. Here to the OSA is “described in terms of their function, information, resource, and organizational aspects. This should be designed with structured engineering methods and made operational in a modular and evolutionary architecture for operational use”.

Application

There are multiple areas of usage:

- In Industrial and Production engineering
- In mechanical engineering
- In electronic design automation (printed circuit board (PCB) and integrated circuit design data for manufacturing)

Computer-Aided Design

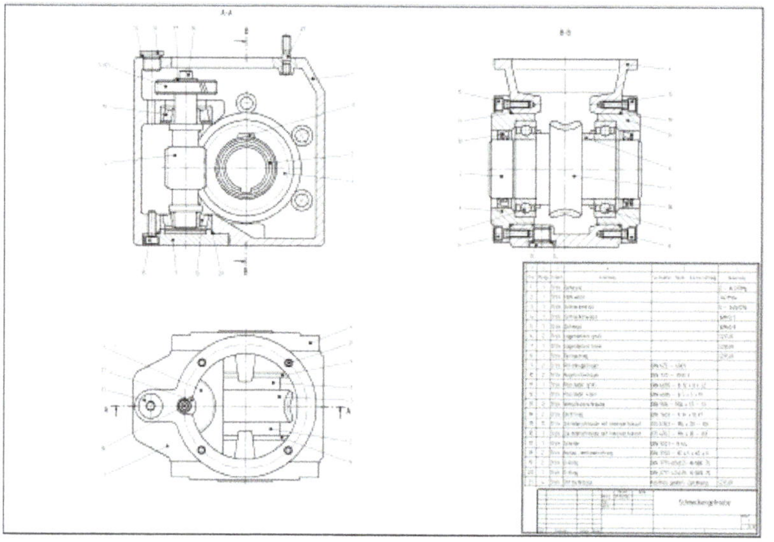

2D CAD drawing

3D CAD model

Computer-aided design (CAD) is the use of computer systems to aid in the creation, modification, analysis, or optimization of a design. CAD software is used to increase the productivity of the designer, improve the quality of design, improve communications through documentation, and to create a database for manufacturing. CAD output is often in the form of electronic files for print, machining, or other manufacturing operations. The term CADD (for *Computer Aided Design and Drafting*) is also used.

Its use in designing electronic systems is known as electronic design automation, or EDA. In mechanical design it is known as mechanical design automation (MDA) or computer-aided drafting (CAD), which includes the process of creating a technical drawing with the use of computer software.

CAD software for mechanical design uses either vector-based graphics to depict the objects of traditional drafting, or may also produce raster graphics showing the overall appearance of designed objects. However, it involves more than just shapes. As in the manual drafting of technical and engineering drawings, the output of CAD must convey information, such as materials, processes, dimensions, and tolerances, according to application-specific conventions.

CAD may be used to design curves and figures in two-dimensional (2D) space; or curves, surfaces, and solids in three-dimensional (3D) space.

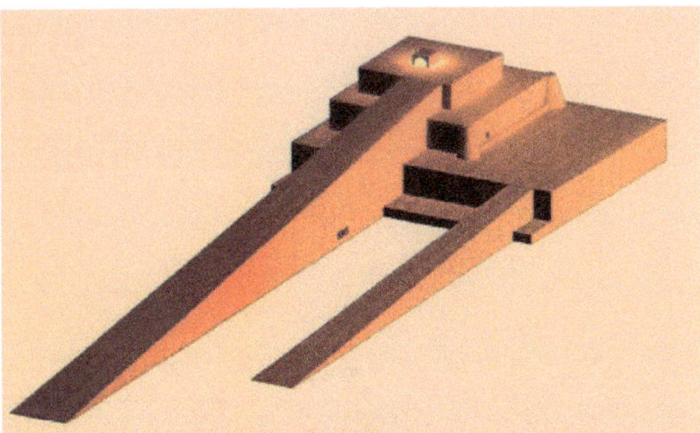

CAD rendering of Sialk ziggurat based on archeological evidence

CAD is an important industrial art extensively used in many applications, including automotive, shipbuilding, and aerospace industries, industrial and architectural design, prosthetics, and many more. CAD is also widely used to produce computer animation for special effects in movies, advertising and technical manuals, often called DCC digital content creation. The modern ubiquity and power of computers means that even perfume bottles and shampoo dispensers are designed using techniques unheard of by engineers of the 1960s. Because of its enormous economic importance, CAD has been a major driving force for research in computational geometry, computer graphics (both hardware and software), and discrete differential geometry.

The design of geometric models for object shapes, in particular, is occasionally called *computer-aided geometric design (CAGD)*.

Overview of CAD Software

Starting around the mid 1970s, as computer aided design systems began to provide more capability than just an ability to reproduce manual drafting with electronic drafting, the cost benefit for companies to switch to CAD became apparent. The benefit of CAD systems over manual drafting are the capabilities one often takes for granted from computer systems today; automated generation of Bill of Material, auto layout in integrated circuits, interference checking, and many others. Eventually CAD provided the designer with the ability to perform engineering calculations. During this transition, calculations were still performed either by hand or by those individuals who could run computer programs. CAD was a revolutionary change in the engineering industry, where draftsmen, designers and engineering roles begin to merge. It did not eliminate departments, as much as it merged departments and empowered draftsman, designers and engineers. CAD is just another example of the pervasive effect computers were beginning to have on industry. Current computer-aided design software packages range from 2D vector-based drafting systems to 3D solid and surface modelers. Modern CAD packages can also frequently allow rotations in three dimensions, allowing viewing of a designed object from any desired angle, even from the inside looking out. Some CAD software is capable of dynamic mathematical modeling, in which case it may be marketed as CAD.

CAD technology is used in the design of tools and machinery and in the drafting and design of all types of buildings, from small residential types (houses) to the largest commercial and industrial structures (hospitals and factories).

CAD is mainly used for detailed engineering of 3D models and/or 2D drawings of physical components, but it is also used throughout the engineering process from conceptual design and layout of products, through strength and dynamic analysis of assemblies to definition of manufacturing methods of components. It can also be used to design objects. Furthermore, many CAD applications now offer advanced rendering and animation capabilities so engineers can better visualize their product designs. 4D BIM is a type of virtual construction engineering simulation incorporating time or schedule related information for project management.

CAD has become an especially important technology within the scope of computer-aided technologies, with benefits such as lower product development costs and a greatly shortened design cycle. CAD enables designers to layout and develop work on screen, print it out and save it for future editing, saving time on their drawings.

Uses

Computer-aided design is one of the many tools used by engineers and designers and is used in many ways depending on the profession of the user and the type of software in question.

CAD is one part of the whole Digital Product Development (DPD) activity within the Product Lifecycle Management (PLM) processes, and as such is used together with other tools, which are either integrated modules or stand-alone products, such as:

- Computer-aided engineering (CAE) and Finite element analysis (FEA)
- Computer-aided manufacturing (CAM) including instructions to Computer Numerical Control (CNC) machines
- Photorealistic rendering and Motion Simulation.
- Document management and revision control using Product Data Management (PDM).

CAD is also used for the accurate creation of photo simulations that are often required in the preparation of Environmental Impact Reports, in which computer-aided designs of intended buildings are superimposed into photographs of existing environments to represent what that locale will be like, where the proposed facilities are allowed to be built. Potential blockage of view corridors and shadow studies are also frequently analyzed through the use of CAD.

CAD has been proven to be useful to engineers as well. Using four properties which are history, features, parameterization, and high level constraints. The construction history can be used to look back into the model's personal features and work on the single area rather than the whole model. Parameters and constraints can be used to determine the size, shape, and other properties of the different modeling elements. The features in the CAD system can be used for the variety of tools for measurement such as tensile strength, yield strength, electrical or electro-magnetic properties. Also its stress, strain, timing or how the element gets affected in certain temperatures, etc.

Types

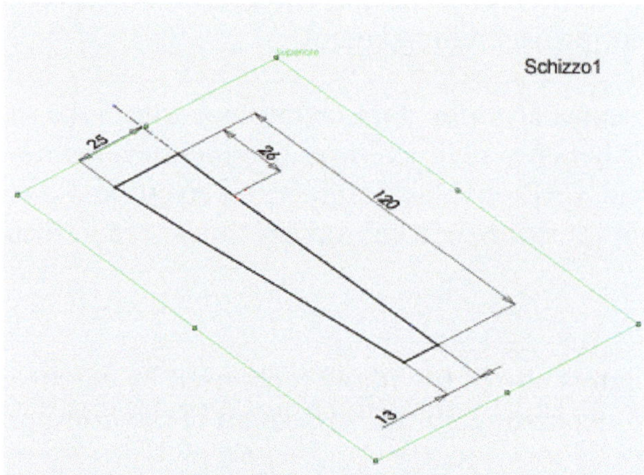

A simple procedure

There are several different types of CAD, each requiring the operator to think differently about how to use them and design their virtual components in a different manner for each.

There are many producers of the lower-end 2D systems, including a number of free and open source programs. These provide an approach to the drawing process without all the fuss over scale and placement on the drawing sheet that accompanied hand drafting, since these can be adjusted as required during the creation of the final draft.

3D wireframe is basically an extension of 2D drafting (not often used today). Each line has to be manually inserted into the drawing. The final product has no mass properties associated with it and cannot have features directly added to it, such as holes. The operator approaches these in a similar fashion to the 2D systems, although many 3D systems allow using the wireframe model to make the final engineering drawing views.

3D "dumb" solids are created in a way analogous to manipulations of real world objects (not often used today). Basic three-dimensional geometric forms (prisms, cylinders, spheres, and so on) have solid volumes added or subtracted from them, as if assembling or cutting real-world objects. Two-dimensional projected views can easily be generated from the models. Basic 3D solids don't usually include tools to easily allow motion of components, set limits to their motion, or identify interference between components.

There are two types of *3D Solid Modeling*

1. *Parametric modeling* allows the operator to use what is referred to as "design intent". The objects and features created are modifiable. Any future modifications can be made by changing how the original part was created. If a feature was intended to be located from the center of the part, the operator should locate it from the center of the model. The feature could be located using any geo-

metric object already available in the part, but this random placement would defeat the design intent. If the operator designs the part as it functions the parametric modeler is able to make changes to the part while maintaining geometric and functional relationships.

2. *Direct or Explicit modeling* provide the ability to edit geometry without a history tree. With direct modeling once a sketch is used to create geometry the sketch is incorporated into the new geometry and the designer just modifies the geometry without needing the original sketch. As with parametric modeling, direct modeling has the ability to include relationships between selected geometry (e.g., tangency, concentricity).

Top end systems offer the capabilities to incorporate more organic, aesthetics and ergonomic features into designs. Freeform surface modeling is often combined with solids to allow the designer to create products that fit the human form and visual requirements as well as they interface with the machine.

Technology

A CAD model of a computer mouse.

Originally software for Computer-Aided Design systems was developed with computer languages such as Fortran, ALGOL but with the advancement of object-oriented programming methods this has radically changed. Typical modern parametric feature based modeler and freeform surface systems are built around a number of key C modules with their own APIs. A CAD system can be seen as built up from the interaction of a graphical user interface (GUI) with NURBS geometry and/or boundary representation (B-rep) data via a geometric modeling kernel. A geometry constraint engine may also be employed to manage the associative relationships between geometry, such as wireframe geometry in a sketch or components in an assembly.

Unexpected capabilities of these associative relationships have led to a new form of prototyping called digital prototyping. In contrast to physical prototypes, which entail manufacturing time in the design. That said, CAD models can be generated by a computer after the physical prototype has been scanned using an industrial CT scanning machine. Depending on the nature of the business, digital or physical prototypes can be initially chosen according to specific needs.

Today, CAD systems exist for all the major platforms (Windows, Linux, UNIX and Mac OS X); some packages even support multiple platforms.

Right now, no special hardware is required for most CAD software. However, some CAD systems can do graphically and computationally intensive tasks, so a modern graphics card, high speed (and possibly multiple) CPUs and large amounts of RAM may be recommended.

The human-machine interface is generally via a computer mouse but can also be via a pen and digitizing graphics tablet. Manipulation of the view of the model on the screen is also sometimes done with the use of a Spacemouse/SpaceBall. Some systems also support stereoscopic glasses for viewing the 3D model.Technologies which in the past were limited to larger installations or specialist applications have become available to a wide group of users. These include the CAVE or HMD`s and interactive devices like motion-sensing technology

Software

CAD software enables engineers and architects to design, inspect and manage engineering projects within an integrated graphical user interface (GUI) on a personal computer system. Most applications support solid modeling with boundary representation (B-Rep) and NURBS geometry, and enable the same to be published in a variety of formats. A geometric modeling kernel is a software component that provides solid modeling and surface modeling features to CAD applications.

Based on market statistics, commercial software from Autodesk, Dassault Systems, Siemens PLM Software and PTC dominate the CAD industry. The following is a list of major CAD applications, grouped by usage statistics.

Commercial

- ASCON KOMPAS-3D
- Autodesk AutoCAD
- Autodesk Inventor
- Dassault CATIA
- Dassault SolidWorks
- Kubotek KeyCreator
- Siemens NX
- Siemens Solid Edge

- PTC Pro/ENGINEER (now renamed Creo)
- TurboCAD
- IronCAD
- MEDUSA
- ProgeCAD
- SpaceClaim
- Rhinoceros 3D
- VariCAD
- VectorWorks
- Cobalt

Freeware and open source

- 123D
- LibreCAD
- FreeCAD
- BRL-CAD
- OpenSCAD
- NanoCAD
- QCad

CAD Kernels

- Parasolid by Siemens
- ACIS by Spatial
- ShapeManager by Autodesk
- Open CASCADE
- C3D by C3D Labs
- K3 by Center GeoS

History

Designers have long used computers for their calculations. Digital computers were used in power system analysis or optimization as early as proto-"Whirlwind" in 1949. Circuit design theory, or power network methodology would be algebraic, symbolic, and often vector-based. Examples of problems being solved in the mid-1940s to 50s include, Servo motors controlled by generated pulse (1949), The digital computer with built-in compute operations to automatically co-ordinate transforms to compute radar related vectors (1951) and the essentially graphic mathematical process of forming a shape with a digital machine tool (1952). These were accomplished with the use of computer software. The man credited with coining the term CAD. Douglas T. Ross stated "As soon as I saw the interactive display equipment," [being used by radar operators 1953] it would be just what his data reduction group needed. With the Lincoln Lab people, they were the only ones who used the big,complex display systems put in for the pre-SAGE,Cape Cod system. But "we used it for our own personal workstation.". The designers of these very early computers built utility programs so that programmers could debug programs using flow charts on a display scope with logical switches that could be opened and closed during the debugging session. They found that they could create electronic symbols and geometric figures to be used to create simple circuit diagrams and flow charts. They made the pleasant discovery that an object once drawn could be reproduced at will, its orientation, Linkage [flux, mechanical, lexical scoping] or scale changed. This suggested numerous possibilities to them. It took ten years of interdisciplinary development work before SKETCHPAD sitting on evolving math libraries emerged from MIT`s labs. Additional developments were carried out in the 1960s within the aircraft, automotive, industrial control and electronics industries in the area of 3D surface construction, NC programming and design analysis, most of it independent of one another and often not publicly published until much later. Some of the mathematical description work on curves was developed in the early 1940s by Robert Issac Newton from Pawtucket, Rhode Island. Robert A. Heinlein in his 1957 novel *The Door into Summer* suggested the possibility of a robotic *Drafting Dan*. However, probably the most important work on polynomial curves and sculptured surface was done by Pierre Bézier, Paul de Casteljau (Citroen), Steven Anson Coons (MIT, Ford), James Ferguson (Boeing), Carl de Boor (GM), Birkhoff (GM) and Garibedian (GM) in the 1960s and W. Gordon (GM) and R. Riesenfeld in the 1970s.

The invention of the 3D CAD/CAM is attributed to a French engineer, Pierre Bezier (Arts et Métiers ParisTech, Renault). After his mathematical work concerning surfaces, he developed UNISURF, between 1966 and 1968, to ease the design of parts and tools for the automotive industry. Then, UNISURF became the working base for the following generations of CAD software.

It is argued that a turning point was the development of the SKETCHPAD system at MIT by Ivan Sutherland (who later created a graphics technology company with Dr. David Evans). The distinctive feature of SKETCHPAD was that it allowed the designer to interact with his computer graphically: the design can be fed into the computer by drawing on a CRT monitor with a light pen. Effectively, it was a prototype of graphical user interface, an indispensable feature of modern CAD. Sutherland presented his paper Sketchpad: A Man-Machine Graphical Communication System in 1963 at a Joint Computer Conference having worked on it as his PhD thesis paper for a few years. Quoting,"For drawings where motion of the drawing, or analysis of a drawn problem is of value to the user, Sketchpad excels. For highly repetitive drawings or drawings where accuracy is required, Sketchpad is sufficiently faster than conventional techniques to be worthwhile. For drawings which merely communicate with shops, it is probably better to use conventional paper and pencil." Over time efforts would be directed toward the goal of having the designers drawings communicate not just with shops but with the shop tool itself. This goal would be a long time arriving.

The first commercial applications of CAD were in large companies in the automotive and aerospace industries, as well as in electronics. Only large corporations could afford the computers capable of performing the calculations. Notable company projects were, a joint project of GM (Dr. Patrick J.Hanratty) and IBM (Sam Matsa, Doug Ross`s MIT APT research assistant) to develop a prototype system for design engineers DAC-1 (Design Augmented by Computer) 1964; Lockheed projects; Bell GRAPHIC 1 and Renault.

One of the most influential events in the development of CAD was the founding of MCS (Manufacturing and Consulting Services Inc.) in 1971 by Dr. P. J. Hanratty, who wrote the system ADAM (Automated Drafting And Machining) but more importantly supplied code to companies such as McDonnell Douglas (Unigraphics), Computervision (CADDS), Calma, Gerber, Autotrol and Control Data.

As computers became more affordable, the application areas have gradually expanded. The development of CAD software for personal desktop computers was the impetus for almost universal application in all areas of construction.

Other key points in the 1960s and 1970s would be the foundation of CAD systems United Computing, Intergraph, IBM, Intergraph IGDS in 1974 (which led to Bentley Systems MicroStation in 1984).

CAD implementations have evolved dramatically since then. Initially, with 3D in the 1970s, it was typically limited to producing drawings similar to hand-drafted drawings. Advances in programming and computer hardware, notably solid modeling in the 1980s, have allowed more versatile applications of computers in design activities.

Key products for 1981 were the solid modelling packages - Romulus (ShapeData) and Uni-Solid (Unigraphics) based on PADL-2 and the release of the surface modeler CATIA (Dassault Systemes). Autodesk was founded 1982 by John Walker, which led to the 2D system AutoCAD. The next milestone was the release of Pro/ENGINEER in 1987, which heralded greater usage of feature-based modeling methods and parametric linking of the parameters of features. Also of importance to the development of CAD was the development of the B-rep solid modeling kernels (engines for manipulating geometrically and topologically consistent 3D objects) Parasolid (ShapeData) and ACIS (Spatial Technology Inc.) at the end of the 1980s and beginning of the 1990s, both inspired by the work of Ian Braid. This led to the release of mid-range packages such as SolidWorks and TriSpective (later known as IRONCAD) in 1995, Solid Edge (then Intergraph) in 1996 and Autodesk Inventor in 1999. An independent geometric modeling kernel has been evolving in Russia since the 1990s. Nikolay Golovanov joined ASCON Company in 1994 from the Kolomna Engineering Design Bureau and began development of C3D – the geometric kernel of the Russian popular CAD system, KOMPAS-3D. Nowadays, C3D (C3D Labs) is the most valued Russian CAD product in the category of "components", i.e. products designed for integration in the end-user CAD systems of Russian and global vendors.

Mobile Industrial Robots

Mobile industrial robots are pieces of machinery that are able to be programmed to perform tasks in an industrial setting. Typically these have been used in stationary and workbench applications; however, mobile industrial robots introduce a new method for lean manufacturing. With advances in controls and robotics, current technology has been improved allowing for mobile tasks such as product delivery. This additional flexibility in manufacturing can save a company time and money during the manufacturing process, and therefore results in a cheaper end product.

Mobile robot technology has potential to revolutionize many sectors of industry; however, it carries with it some disadvantages. The logistics of manufacturing will be streamlined by allowing robots to autonomously navigate to different areas for their work. The labor demands for employees will be lessened as robots will be able to work alongside humans, and robots will assist with medicine and surgery more and more. However, there are drawbacks to this technology. Coordinating the movement of robots around facilities and calibrating their position at their destination is tedious and far from perfect. A robot malfunctioning in a manufacturing setting will hold up production - and this robot could malfunction anywhere in a facility. Human safety must also be considered. Robots must prioritize the safety of human operators over their programmed task - which may complicate the coordination of multiple autonomous robots. Especially in a surgical setting, there is no room for error on the robot's part. Even though some challenges are present, mobile robot technology promises to streamline aspects across much of the industry.

History

Automation began in the automobile industry in the years surrounding WWII (1946) and the origin of the term itself belongs with D.S. Harder, the engineering manager at the Ford Motor Company. At first, the term was used to describe the increased presence of automatic devices in production lines and solely manufacturing contexts. Now, automation is widely used in many industries where computerized action and feedback loops can replace human intervention in the workplace. Over time, development in this area has become increasingly dependent upon advanced computer technologies and the advancement of processing capabilities.

In its current form, most industrial robots are powered mechanical arms with the ability to perform anthropomorphic actions. Advancements in miniaturization of computers, mathematical control theory as well as improved sensory technologies have had great impact on the feedback control systems that drive robotics. The first industrial robot performed spot welding and die castings in a General Motors factory in New Jersey, USA in 1962. Soon, robotic arms were exploding within the large-scale manufacturing industry and several new companies came into existence including Kuka in 1973, Nachi in 1969, Fanuc in 1974, Yaskawa in 1977, ASEA in 1977, and several others. By 1980, it is estimated a new major robotics company entered the market every month.

Mobile robotics are now set to experience similar expansion as they become significantly more reliable in an industrial setting. Even if a mobile robot makes mistakes, it will eventually be less frequently than mistakes caused by human factors.

Overview

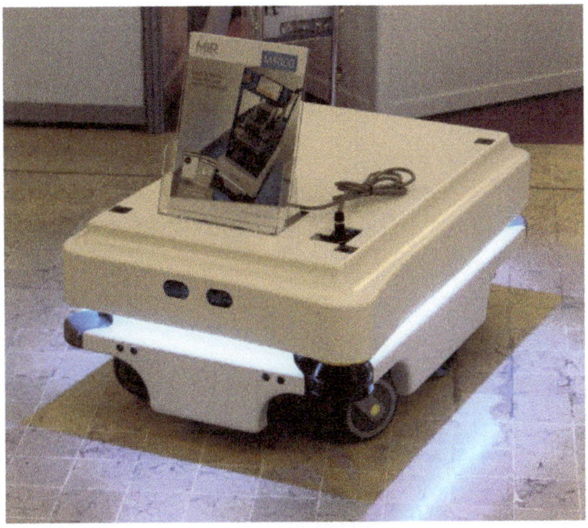

MiR100, a frontrunning example product for mobile industrial robots.

The simplicity of mobile industrial robots provide their main advantage in industrial settings due to the ease of use and ability to be operated via technologies well understood by most people. In addition, robots are able to operate almost continuously and will never complain about long work hours; greatly increasing efficiency in a lean manufacturing environment. The main current disadvantage lies in high costs of repair as well as the production delays that would be caused by a failure or malfunction. These factors are very preventative to putting major amounts of responsibility on mobile robotics, however they are being continually lessened.

Applications of Mobile Industrial Robots

The mobile industrial robots have many applications that they have been used in already including in the healthcare industry, home and industrial security, ocean and space exploration, the food service industry, and in distribution applications.

Medicine

Mobile industrial robots have several uses within the healthcare industry in both hospitals and homes. Drug delivery, patient services, and other nursing functions could be easily adapted to robots. Due to the fact that items being carried around typically weigh less than 100 kg, robots much smaller than the MiR may be used. Specialized equipment may be mounted on robots, allowing them to assist with surgical procedures. Overall, their place in the medical industry is to provide a more reliable source of customer care while reducing human error.

Scientific Experimentation and Exploration

In the scientific world, there is a large number of applications for mobile robots. Their ability to perform experiments and exploration without putting human lives in danger makes them an important asset. Unlike humans, robots do not require life support systems to function. In space travel, robots are performing science on planets and asteroids because sending humans is far more taxing on resources and money. The same is true in the oceanography domain. In fact, several of the same robotic systems are designed to perform their science under both conditions - space and underwater. In nuclear power plants, robots can service electronics and mechanical systems which prevents human exposure to large amounts of radiation.

Aircraft Maintenance and Repair

For applications like painting and de-painting aircraft, two fixed robots are inadequate because not all parts of the aircraft can be reached. Adding more fixed robots would complete the task, but the cost is prohibitive. If mobile robots are used, one or two may be enough to service the entire aircraft because they can move to whatever area needs work. Mobile robots need to be truly autonomous to be useful in manufacturing. Erik Nieves said "Mobility moves robots from being machines to production partners [...]" Rather than bringing work to the robot, the robot should be smart enough to go to where the work is.

Automated aircraft inspection systems have the potential to make aircraft maintenance safer and more reliable. Various solutions are currently developed: a collaborative mobile robot named Air-Cobot, and drones.

Air-Cobot is a collaborative mobile robot able to inspect aircraft. Picture of the robot in Air France Industries.

Examples

OTTO Motors (A Division of Clearpath)

OTTO 1500 self-driving vehicle for heavy-load material transport in warehouses, distribution centers, and factories.

Meant for material transport in industrial centers

- Able to carry 100 kg (OTTO 100) or 1500 kg (OTTO 1500)
- Powered by lithium battery technology
- 6-8 hours operating time (depending on payload)
- Requires no structural changes to building
- Navigates via 2D sensors. Proprietary autonomy software enables dynamic path-planning and obstacle avoidance at speeds matching forklifts (2 m/s).
- Used by GE and John Deere.

MiR (Mobile Industrial Robots Corp.)

KUKA youBot

- Meant for transportation
  - Able to carry 100 kg, large enough to hold full size standard EU pallets
  - Electrically powered
  - Long operating time on a single battery charge (12–15 hours/20 km)
  - Maximum velocity: 1.5 m/s
- Requires no structural changes of building (i.e. wires in the floor, sensors in the ceiling)
- Can be operated via smartphone or PC
- Navigates via an array of 2D and 3D scanners and cameras

Kuka

- Very widely used—Example: Tesla Motors
- "Mecanum" wheel system: customizeable, modular, heavy-lifting capable
- Very easy to integrate with autonomous robotics and humans

Pilot Plant

A pilot plant is a small industrial system, which is operated to generate information about the behavior of the system for use in design of larger facilities. Pilot plant is a relative term in the sense that plants are typically smaller than full-scale production plants, but are built in a range of sizes. Some pilot plants are built in laboratories using stock lab equipment, while others require substantial engineering efforts, cost millions of dollars, and are custom-assembled and fabricated from process equipment, instrumentation and piping. They can also be used to train personnel for a full-scale plant.

Pilot plants tend to be smaller compared to demonstration plants.

A pilot plant of a bioreactor

Risk Management

Pilot plants are used to reduce the risk associated with construction of large process plants. They do so in several ways:

- Computer simulations and semi-empirical methods are used to determine the limitations of the pilot scale system. These mathematical models are then tested in a physical pilot-scale plant. Various modeling methods are used for scale-up. These methods include:
    - Chemical similitude studies
    - Mathematical modeling
    - Aspen Plus/Aspen HYSYS modeling
    - Finite Elemental Analysis (FEA)
    - Computational Fluid Dynamics (CFD)
        - These theoretical modeling methods return the following:
            - Finalized mass and energy balances
            - Optimized system design and capacity
            - Equipment requirements

   - System limitations
   - The basis for determining the cost to build the pilot module

- They are substantially less expensive to build than full-scale plants. The business does not put as much capital at risk on a project that may be inefficient or unfeasible. Further, design changes can be made more cheaply at the pilot scale and kinks in the process can be worked out before the large plant is constructed.
- They provide valuable data for design of the full-scale plant. Scientific data about reactions, material properties, corrosiveness, for instance, may be available, but it is difficult to predict the behavior of a process of any complexity. Engineering data from other process may be available, but this data can not always be clearly applied to the process of interest. Designers use data from the pilot plant to refine their design of the production scale facility.

If a system is well defined and the engineering parameters are known, pilot plants are not used. For instance, a business that wants to expand production capacity by building a new plant that does the same thing as an existing plant may choose to not use a pilot plant.

Additionally, advances in process simulation on computers have increased the confidence of process designers and reduced the need for pilot plants. However, they are still used as even state-of-the-art simulation cannot accurately predict the behavior of complex systems.

Design Change

As a system increases in size, system properties that depend on quantity of matter (with extensive properties) might change. The chemical and physical properties of a system affect each other and create varying results. The surface area to liquid ratio is a good example of such a property. On a small chemical scale, in a flask, say, there is a relatively large surface area to liquid ratio. However, if the reaction in question is scaled up to fit in a 500-gallon tank, the surface area to liquid ratio becomes much smaller. As a result of this difference in surface area to liquid ratio, the exact nature of the thermodynamics and the reaction kinetics of the process changes in a non-linear fashion. This is why a reaction in a beaker can behave vastly differently from the same reaction in a large-scale production process.

Other Factors

Other factors that change during the transformation to a production scale include:

- Reaction kinetics

- Chemical equilibrium
- Material properties
- Fluid dynamics
- Thermodynamics
- Equipment selection
- Agitation

After data is collected from operation of a pilot plant, a larger production-scale facility may be built. Alternatively, a demonstration plant, which is bigger than a pilot plant, but smaller than the full-scale production plant, may be built to demonstrate the commercial feasibility of the process. Businesses sometimes continue to operate the pilot plant in order to test ideas for new products, new feedstocks, or different operating conditions. Alternatively, they may be operated as production facilities, augmenting production from the main plant.

Recent trends try to keep the size of the plant a small as possible to save costs. This approach is called miniplant technology. The flow chemistry takes up this trend and uses flow miniplant technology for small-scale manufacturing.

Bench Scale vs Pilot vs Demonstration

The differences between bench scale, pilot scale and demonstration scale are strongly influenced by industry and application. Some industries use pilot plant and demonstration plant interchangeably. Some pilot plants are built as portable modules that can easily transported as a contained unit.

For batch processes, in the pharmaceutical industry for example, bench scale is typically conducted on samples 1–20 kg or less, whereas pilot scale testing is performed with samples of 20–100 kg. Demonstration scale is essentially operating the equipment at full commercial feed rates over extended time periods to prove operational stability.

For continuous processes, in the petroleum industry for example, bench scale systems are typically microreactor or CSTR systems with less than 1000 cc of catalyst, studying reactions and/or separations on a once-through basis. Pilot plants will typically have reactors with catalyst volume between 1 and 100 litres filled with Jesse, and will often incorporate product separation and gas/liquid recycle with the goal of closing the mass balance. Demonstration plants, also referred to as semi-works plants, will study the viability of the process on a pre-commercial scale, with typical catalyst volumes in the 100 - 1000 litre range. The design of a demonstration scale plant for a continuous process will closely resemble that of the anticipated future commercial plant, albeit at a much lower throughput, and its goal is to study catalyst performance and operating lifetime over an extended period, while generating significant quantities of product for market testing.

In the development of new processes, the design and operation of the pilot and demonstration plant will often run in parallel with the design of the future commercial plant, and the results from pilot testing programs are key to optimizing the commercial plant flowsheet. It is common in cases where process technology has been successfully implemented that the savings at the commercial scale resulting from pilot testing will significantly outweigh the cost of the pilot plant itself.

Steps to Creating a Custom Pilot Plant

Custom pilot plants are commonly designed either for research or commercial purposes. They can range in size from a small system with no automation and low flow, to a highly automated system producing relatively large amounts of products in a day. No matter the size, the steps to designing and fabricating a working pilot plant are the same. They are:

1. Pre-engineering - completing a process flow diagram (PFD), basic piping and instrumentation diagrams (P&ID's) and initial equipment layouts.

2. Engineering modeling and optimization - 2D and 3D models are created, using a simulation software to model the process parameters and scale the chemical processes. These modeling softwares help determine system limitations, non-linear chemical and physical changes, and potential equipment sizing. Mass and energy balances, finalized P&ID's and general arrangement drawings are produced.

3. Automation strategies for the system are developed (if needed). Controls system programming begins and will continue through fabrication and assembly

4. Fabrication and assembly - after an optimized design has been determined, the custom pilot is fabricated and assembled. Pilot plants can either be assembled on-site or off-site as modular skids that will be constructed and tested in a controlled environment.

5. Testing - testing of completed systems, including system controls, is conducted to ensure proper system function.

6. Installation and startup - if constructed offsite, pilot skids are installed onsite. After all equipment is in place, full system startup is completed by integrating the system with existing plant utilities and controls. Full operation is tested and affirmed.

7. Training - operator training is complete and full system documentation is handed over.

References

- Anna Clark (1997), The struggle for the breeches: gender and the making of the British working class, University of California Press, p. 32ff, ISBN 0520208838
- Geoffrey Timmins (1993), The last shift: the decline of handloom weaving in nineteenth-century Lancashire, Manchester University Press ND, p. 19ff, ISBN 0719037255
- Gail Fowler Mohanty (2006), Labor and laborers of the loom: mechanization and handloom weavers, 1780-1840, CRC Press, p. 114ff, ISBN 0415979021
- Neil J. Smelser (2006), Social Change in the Industrial Revolution: An Application of Theory to the British Cotton Industry, Taylor & Francis, pp. 208–209, ISBN 0415381371
- Rifkin, Jeremy (1995). The End of Work: The Decline of the Global Labor Force and the Dawn of the Post-Market Era. Putnam Publishing Group. pp. 66, 75. ISBN 0-87477-779-8.
- Bennett, S. (1979). A History of Control Engineering 1800-1930. London: Peter Peregrinus Ltd. pp. 47, 266. ISBN 0-86341-047-2.
- Field, Alexander J. (2011). A Great Leap Forward: 1930s Depression and U.S. Economic Growth. New Haven, London: Yale University Press. ISBN 978-0-300-15109-1.
- Constable, George; Somerville, Bob (1964). A Century of Innovation: Twenty Engineering Achievements That Transformed Our Lives. Joseph Henry Press. ISBN 0309089085.
- Pearce, Joshua M. (2014-01-01). Chapter 1 - Introduction to Open-Source Hardware for Science. Boston: Elsevier. pp. 1–11. doi:10.1016/b978-0-12-410462-4.00001-9. ISBN 9780124104624.
- U.S. Congress, Office of Technology Assessment (1984). Computerized manufacturing automation. DIANE Publishing. p. 48. ISBN 978-1-4289-2364-5.
- Hosking, Dian Marie; Anderson, Neil (1992), Organizational change and innovation, Taylor & Francis, p. 240, ISBN 978-0-415-06314-2
- Daintith, John (2004). A dictionary of computing (5 ed.). Oxford University Press. p. 102. ISBN 978-0-19-860877-6.
- Kreith, Frank (1998). The CRC handbook of mechanical engineering. CRC Press. p. 15-1. ISBN 978-0-8493-9418-8.
- Matthews, Clifford (2005). Aeronautical engineer's data book (2nd ed.). Butterworth-Heinemann. p. 229. ISBN 978-0-7506-5125-7.
- Pichler, Franz; Moreno-Díaz, Roberto (1992). Computer aided systems theory. Springer. p. 602. ISBN 978-3-540-55354-0.
- Boothroyd, Geoffrey; Knight, Winston Anthony (2006). Fundamentals of machining and machine tools (3rd ed.). CRC Press. p. 401. ISBN 978-1-57444-659-3
- Kalpakjian, Serope; Schmid, Steven (2006), Manufacturing engineering and technology (5th ed.), Prentice Hall, p. 1192, ISBN 978-7-302-12535-8.
- Laplante, Phillip A. (2005), Comprehensive dictionary of electrical engineering (2nd ed.), CRC Press, p. 136, ISBN 978-0-8493-3086-5.
- Waldner, Jean-Baptiste (September 1992). Principles of Computer-Integrated Manufacturing. London: John Wiley & Sons. pp. 128–p132. ISBN 0-471-93450-X.
- Narayan, K. Lalit (2008). Computer Aided Design and Manufacturing. New Delhi: Prentice Hall of India. p. 3. ISBN 812033342X.

- Narayan, K. Lalit (2008). Computer Aided Design and Manufacturing. New Delhi: Prentice Hall of India. p. 4. ISBN 812033342X.
- Duggal, Vijay (2000). Cadd Primer: A General Guide to Computer Aided Design and Drafting-Cadd, CAD. Mailmax Pub. ISBN 978-0962916595.
- Farin, Gerald; Hoschek, Josef; Kim, Myung-Soo (2002). Handbook of computer aided geometric design [electronic resource]. Elsevier. ISBN 978-0-444-51104-1.
- Pottmann, H.; Brell-Cokcan, S. and Wallner, J. (2007) "Discrete surfaces for architectural design", pp. 213–234 in Curve and Surface Design, Patrick Chenin, Tom Lyche and Larry L. Schumaker (eds.), Nashboro Press, ISBN 978-0-9728482-7-5.

5

Machining and Machine Tools

Machining is the procedure in which raw material is cut into a particular shape and then is controlled by a material removal process. The various processes of machining elucidated in the text are hobbing and milling. The other tools explained in the section are shapers, planers and grinding machines. The chapter closely examines the concept of machining and provides an extensive understanding of the subject.

Machining

New Guinea in 1943. Mobile machine shop truck of the US Army with machinists working on automotive parts

Machining is any of various processes in which a piece of raw material is cut into a desired final shape and size by a controlled material-removal process. The processes that have this common theme, controlled material removal, are today collectively known as subtractive manufacturing, in distinction from processes of controlled material addition, which are known as additive manufacturing. Exactly what the "controlled" part of the definition implies can vary, but it almost always implies the use of machine tools (in addition to just power tools and hand tools).

Machining is a part of the manufacture of many metal products, but it can also be used on materials such as wood, plastic, ceramic, and composites. A person who specializes in machining is called a machinist. A room, building, or company where machining is done is called a machine shop. Machining can be a business, a hobby, or both. Much of modern day machining is carried out by computer numerical control (CNC), in which computers are used to control the movement and operation of the mills, lathes, and other cutting machines.

History and Terminology

The precise meaning of the term *machining* has evolved over the past one and a half centuries as technology has advanced. In the 18th century, the word *machinist* simply meant a person who built or repaired machines. This person's work was done mostly by hand, using processes such as the carving of wood and the hand-forging and hand-filing of metal. At the time, millwrights and builders of new kinds of *engines* (meaning, more or less, machines of any kind), such as James Watt or John Wilkinson, would fit the definition. The noun *machine tool* and the verb *to machine* (*machined, machining*) did not yet exist.

Around the middle of the 19th century, the latter words were coined as the concepts that they described evolved into widespread existence. Therefore, during the Machine Age, *machining* referred to (what we today might call) the "traditional" machining processes, such as turning, boring, drilling, milling, broaching, sawing, shaping, planing, reaming, and tapping. In these "traditional" or "conventional" machining processes, machine tools, such as lathes, milling machines, drill presses, or others, are used with a sharp cutting tool to remove material to achieve a desired geometry.

Since the advent of new technologies such as electrical discharge machining, electrochemical machining, electron beam machining, photochemical machining, and ultrasonic machining, the retronym "conventional machining" can be used to differentiate those classic technologies from the newer ones. In current usage, the term "machining" without qualification usually implies the traditional machining processes.

Machining Operations

Making a shipboard manhole cover in the machine shop of the aircraft carrier USS *John C. Stennis*.

The three principal machining processes are classified as turning, drilling and milling. Other operations falling into miscellaneous categories include shaping, planing, boring, broaching and sawing.

- Turning operations are operations that rotate the workpiece as the primary method of moving metal against the cutting tool. Lathes are the principal machine tool used in turning.
- Milling operations are operations in which the cutting tool rotates to bring cut-

ting edges to bear against the workpiece. Milling machines are the principal machine tool used in milling.

- Drilling operations are operations in which holes are produced or refined by bringing a rotating cutter with cutting edges at the lower extremity into contact with the workpiece. Drilling operations are done primarily in drill presses but sometimes on lathes or mills.
- Miscellaneous operations are operations that strictly speaking may not be machining operations in that they may not be swarf producing operations but these operations are performed at a typical machine tool. Burnishing is an example of a miscellaneous operation. Burnishing produces no swarf but can be performed at a lathe, mill, or drill press.

An unfinished workpiece requiring machining will need to have some material cut away to create a finished product. A finished product would be a workpiece that meets the specifications set out for that workpiece by engineering drawings or blueprints. For example, a workpiece may be required to have a specific outside diameter. A lathe is a machine tool that can be used to create that diameter by rotating a metal workpiece, so that a cutting tool can cut metal away, creating a smooth, round surface matching the required diameter and surface finish. A drill can be used to remove metal in the shape of a cylindrical hole. Other tools that may be used for various types of metal removal are milling machines, saws, and grinding machines. Many of these same techniques are used in woodworking.

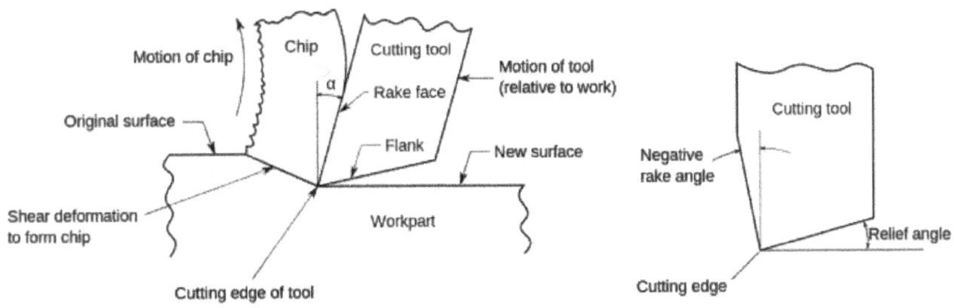

Basic machining process.

More recent, advanced machining techniques include precision CNC machining, electrical discharge machining (EDM), electro-chemical erosion, laser cutting, or water jet cutting to shape metal workpieces.

As a commercial venture, machining is generally performed in a machine shop, which consists of one or more workrooms containing major machine tools. Although a machine shop can be a stand-alone operation, many businesses maintain internal machine shops which support specialized needs of the business.

Machining requires attention to many details for a workpiece to meet the specifications set out in the engineering drawings or blueprints. Beside the obvious problems related to correct dimensions, there is the problem of achieving the correct finish or surface smoothness on the workpiece. The inferior finish found on the machined surface of a workpiece may be caused by incorrect clamping, a dull tool, or inappropriate presentation of a tool. Frequently, this poor surface finish, known as chatter, is evident by an undulating or irregular finish, and the appearance of waves on the machined surfaces of the workpiece.

Overview of Machining Technology

Machining is any process in which a cutting tool is used to remove small chips of material from the workpiece (the workpiece is often called the "work"). To perform the operation, relative motion is required between the tool and the work. This relative motion is achieved in most machining operation by means of a primary motion, called "cutting speed" and a secondary motion called "feed". The shape of the tool and its penetration into the work surface, combined with these motions, produce the desired shape of the resulting work surface.

Machining Operations

There are many kinds of machining operations, each of which is capable of generating a certain part geometry and surface texture.

In turning, a cutting tool with a single cutting edge is used to remove material from a rotating workpiece to generate a cylindrical shape. The primary motion is provided by rotating the workpiece, and the feed motion is achieved by moving the cutting tool slowly in a direction parallel to the axis of rotation of the workpiece.

Drilling is used to create a round hole. It is accomplished by a rotating tool that typically has two or four helical cutting edges. The tool is fed in a direction parallel to its axis of rotation into the workpiece to form the round hole.

In boring, a tool with a single bent pointed tip is advanced into a roughly made hole in a spinning workpiece to slightly enlarge the hole and improve its accuracy. It is a fine finishing operation used in the final stages of product manufacture.

Reaming, is one of the sizing operation that removes a small amount of metal from a hole that already drilled.

In milling, a rotating tool with multiple cutting edges is moved slowly relative to the material to generate a plane or straight surface. The direction of the feed motion is perpendicular to the tool's axis of rotation. The speed motion is provided by the rotating milling cutter. The two basic forms of milling are:

- Peripheral milling
- Face milling.

Other conventional machining operations include shaping, planing, broaching and sawing. Also, grinding and similar abrasive operations are often included within the category of machining.

The Cutting Tool

A cutting tool has one or more sharp cutting edges and is made of a material that is harder than the work material. The cutting edge serves to separate chip from the parent work material. Connected to the cutting edge are the two surfaces of the tool:

- The rake face; and
- The flank.

A "numerical controlled machining cell machinist" monitors a B-1B aircraft part being manufactured.

The rake face which directs the flow of newly formed chip, is oriented at a certain angle is called the rake angle "α". It is measured relative to the plane perpendicular to the work surface. The rake angle can be positive or negative. The flank of the tool provides a clearance between the tool and the newly formed work surface, thus protecting the surface from abrasion, which would degrade the finish. This angle between the work surface and the flank surface is called the relief angle. There are two basic types of cutting tools:

- Single point tool; and
- Multiple-cutting-edge tool

A single point tool has one cutting edge and is used for turning, boreing and planing. During machining, the point of the tool penetrates below the original work surface of the workpart. The point is sometimes rounded to a certain radius, called the nose radius.

Multiple-cutting-edge tools have more than one cutting edge and usually achieve their motion relative to the workpart by rotating. Drilling and milling uses rotating multiple-cutting-edge tools. Although the shapes of these tools are different from a single-point tool, many elements of tool geometry are similar.

Cutting Conditions

Relative motion is required between the tool and work to perform a machining operation. The primary motion is accomplished at a certain cutting speed. In addition, the tool must be moved laterally across the work. This is a much slower motion, called the feed. The remaining dimension of the cut is the penetration of the cutting tool below the original work surface, called the depth of cut. Collectively, speed, feed, and depth of cut are called the cutting conditions. They form the three dimensions of the machining process, and for certain operations, their product can be used to obtain the material removal rate for the process:

$$R_{MR} = vfd$$

where

- R_{MR} – the material removal rate in mm^3/s, (in^3/s),
- v – the cutting speed in mm/s, (in/min),
- f – the feed in mm, (in),
- d – the depth of cut in mm, (in).

  Note: All units must be converted to the corresponding decimal (or USCU) units.

Stages in Metal Cutting

Machining operations usually divide into two categories, distinguished by purpose and cutting conditions:

- Roughing cuts
- Finishing cuts

Roughing cuts are used to remove large amount of material from the starting workpart as rapidly as possible, i.e. with a large Material Removal Rate (MRR), in order to produce a shape close to the desired form, but leaving some material on the piece for a subsequent finishing operation. Finishing cuts are used to complete the part and achieve the final dimension, tolerances, and surface finish. In production machining jobs, one or more roughing cuts are usually performed on the work, followed by one or two finishing cuts. Roughing operations are done at high feeds and depths – feeds of 0.4–1.25 mm/rev (0.015–0.050 in/rev) and depths of 2.5–20 mm (0.100–0.750 in) are typical, but actual values depend on the workpiece materials. Finishing operations are carried out at low feeds and depths – feeds of 0.0125–0.04 mm/rev (0.0005–0.0015 in/rev) and depths of 0.75–2.0 mm (0.030–0.075 in) are typical. Cutting speeds are lower in roughing than in finishing.

A cutting fluid is often applied to the machining operation to cool and lubricate the cutting tool. Determining whether a cutting fluid should be used, and, if so, choosing the proper cutting fluid, is usually included within the scope of cutting condition.

Today other forms of metal cutting are becoming increasingly popular. An example of this is water jet cutting. Water jet cutting involves pressurized water in excess of 620 MPa (90 000 psi) and is able to cut metal and have a finished product. This process is called cold cutting, which eliminates the damage caused by a heat affected zone, as opposed to laser and plasma cutting.

Relationship of Subtractive and Additive Techniques

With the recent proliferation of additive manufacturing technologies, conventional machining has been retronymously classified, in thought and language, as a subtractive manufacturing method. In narrow contexts, additive and subtractive methods may compete with each other. In the broad context of entire industries, their relationship is complementary. Each method has its own advantages over the other. While additive manufacturing methods can produce very intricate prototype designs impossible to replicate by machining, strength and material selection may be limited.

Various Machining Processes

Hobbing

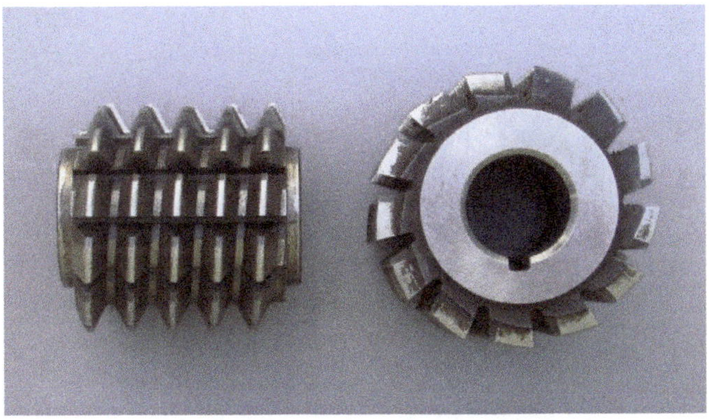

A hob — the cutter used for hobbing.

Hobbing is a machining process for gear cutting, cutting splines, and cutting sprockets on a hobbing machine, which is a special type of milling machine. The teeth or splines are progressively cut into the workpiece by a series of cuts made by a cutting tool called a hob. Compared to other gear forming processes it is relatively inexpensive but still quite accurate, thus it is used for a broad range of parts and quantities.

It is the most widely used gear cutting process for creating spur and helical gears and more gears are cut by hobbing than any other process as it is relatively quick and inexpensive.

A type of skiving that is analogous to the hobbing of external gears can be applied to the cutting of internal gears, which are skived with a rotary cutter (rather than shaped or broached).

Process

Hobbing uses a hobbing machine with two skew spindles, one mounted with a blank workpiece and the other with the hob. The angle between the hob's spindle (axis) and the workpiece's spindle varies, depending on the type of product being produced. For example, if a spur gear is being produced, then the hob is angled equal to the helix angle of the hob; if a helical gear is being produced then the angle must be increased by the same amount as the helix angle of the helical gear. The two shafts are rotated at a proportional ratio, which determines the number of teeth on the blank; for example, for a single-threaded hob if the gear ratio is 40:1 the hob rotates 40 times to each turn of the blank, which produces 40 teeth in the blank. If the hob has multiple threads the speed ratio must be multiplied by the number of threads on the hob. The hob is then fed up into the workpiece until the correct tooth depth is obtained. Finally the hob is fed through the workpiece parallel to the blank's axis of rotation.

Often multiple blanks are stacked, then cut in one operation.

For very large gears the blank can be gashed to the rough shape first to make hobbing easier.

Equipment

A horizontal hobbing machine

Hobbing machines, also known as *hobbers*, are fully automated machines that come in many sizes, because they need to be able to produce anything from tiny instrument gears up to 10 ft (3.0 m) diameter marine gears. Each gear hobbing machine typically consists of a chuck and tailstock, to hold the workpiece or a spindle, a spindle on which the hob is mounted, and a drive motor.

For a tooth profile which is a theoretical involute, the fundamental rack is straight-sided, with sides inclined at the pressure angle of the tooth form, with flat top and bottom. The necessary addendum correction to allow the use of small-numbered pinions can either be obtained by suitable modification of this rack to a cycloidal form at the tips, or by hobbing at other than the theoretical pitch circle diameter. Since the gear ratio between hob and blank is fixed, the resulting gear will have the correct pitch on the pitch circle, but the tooth thickness will not be equal to the space width.

A horizontal hobbing machine

Hobbing machines are characterised by the largest module or pitch diameter it can generate. For example, a 10 in (250 mm) capacity machine can generate gears with a 10 in pitch diameter and usually a maximum of a 10 in face width. Most hobbing machines are vertical hobbers, which means the blank is mounted vertically. Horizontal hobbing machines are usually used for cutting longer workpieces; i.e. cutting splines on the end of a shaft.

Hob

A gear hob in a hobbing machine with a finished gear.

The *hob* is a cutting tool used to cut the teeth into the workpiece. It is cylindrical in shape with helical cutting teeth. These teeth have grooves that run the length of the hob, which aid in cutting and chip removal. There are also special hobs designed for special gears such as the spline and sprocket gears.

The cross-sectional shape of the hob teeth are almost the same shape as teeth of a rack gear that would be used with the finished product. There are slight changes to the shape for generating purposes, such as extending the hob's tooth length to create a clearance in the gear's roots. Each hob tooth is relieved on the back side to reduce friction.

Most hobs are single-thread hobs, but double-, and triple-thread hobs increase production rates. The downside is that they are not as accurate as single-thread hobs. Depending on type of gear teeth to be cut, there are custom made hobs and general purpose hobs. Custom made hobs are different from other hobs as they are suited to make gears with modified tooth profile. The tooth profile is modified to add strength and reduce size and gear noise.

This list outlines types of hobs:

- Roller chain sprocket hobs
- Worm wheel hobs
- Spline hobs
- Chamfer hobs
- Spur and helical gear hobs
- Straight side spline hobs
- Involute spline hobs
- Serration hobs
- Semitopping gear hobs

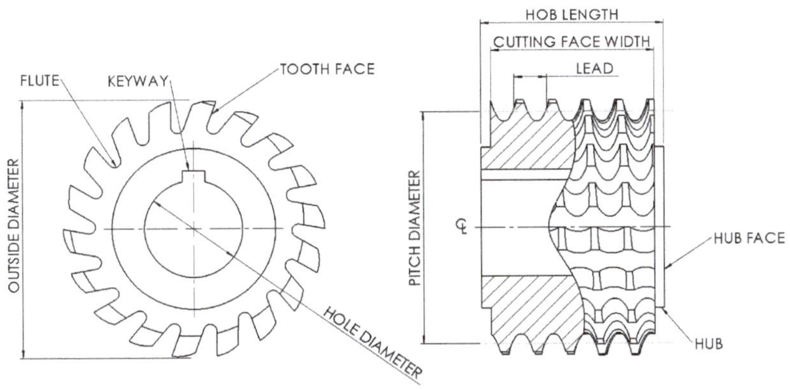

Uses

Hobbing is used to make the following types of finished gears:

- Cycloid gears
- Helical gears
- Involute gears
- Ratchets
- Splines
- Sprockets
- Spur gears
- Worm gears

Hobbing is used to produce most throated worm wheels, but certain tooth profiles cannot be hobbed. If any portion of the hob profile is perpendicular to the axis then it will have no cutting clearance generated by the usual backing off process, and it will not cut well.

Cycloidal Forms

For cycloidal gears (as used in BS978-2 Specification for fine pitch gears) and cycloidal-type gears each module, ratio and number of teeth in the pinion requires a different hobbing cutter, so the technique is only suitable for large volume production.

To circumvent this problem a special war-time emergency circular arc gear standard was produced giving a series of close-to-cycloidal forms which could be cut with a single hob for each module for eight teeth and upwards to economize on cutter manufacturing resources. A variant on this is still included in BS978-2a (Gears for instruments and clockwork mechanisms. Cycloidal type gears. Double circular arc type gears).

Tolerances of concentricity of the hob limit the lower modules which can be cut practically by hobbing to about 0.5 module.

History

Many manufacturing firms that maintain museums of how products were produced in times past will have examples of manual gear hobs that helped to produce gears prior to the gears of the 19th century and earlier. Along with these completely manual gear hobs will be samples of some of the first semi-automated gear hobs, and finally examples of more recent technology that demonstrates the fully automated process that modern gear hobs use to produce gears today. A few producers of gear hobs also have interesting literature on the history of gear hobs, including details about how modern gear hobs can produce thousands of gears in a single hour.

Milling (Machining)

Milling is the machining process of using rotary cutters to remove material from a workpiece by advancing (or feeding) in a direction at an angle with the axis of the tool. It covers a wide variety of different operations and machines, on scales from small individual parts to large, heavy-duty gang milling operations. It is one of the most commonly used processes in industry and machine shops today for machining parts to precise sizes and shapes.

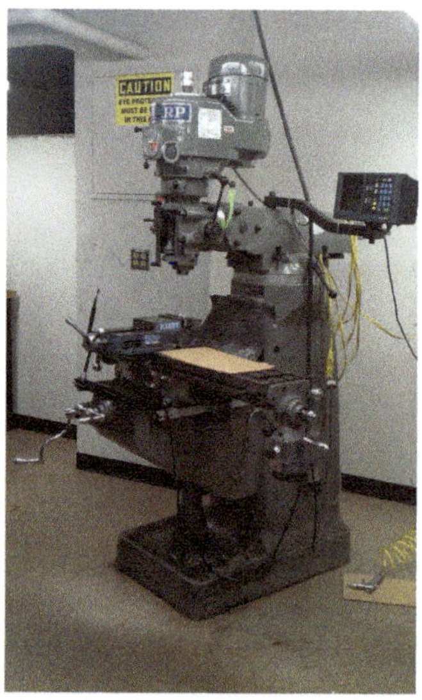

Full view of a Bridgeport clone.

Milling can be done with a wide range of machine tools. The original class of machine tools for milling was the milling machine (often called a mill). After the advent of computer numerical control (CNC), milling machines evolved into machining centers (milling machines with automatic tool changers, tool magazines or carousels, CNC control, coolant systems, and enclosures), generally classified as vertical machining centers (VMCs) and horizontal machining centers (HMCs). The integration of milling into turning environments and of turning into milling environments, begun with live tooling for lathes and the occasional use of mills for turning operations, led to a new class of machine tools, multitasking machines (MTMs), which are purpose-built to provide for a default machining strategy of using any combination of milling and turning within the same work envelope.

Process

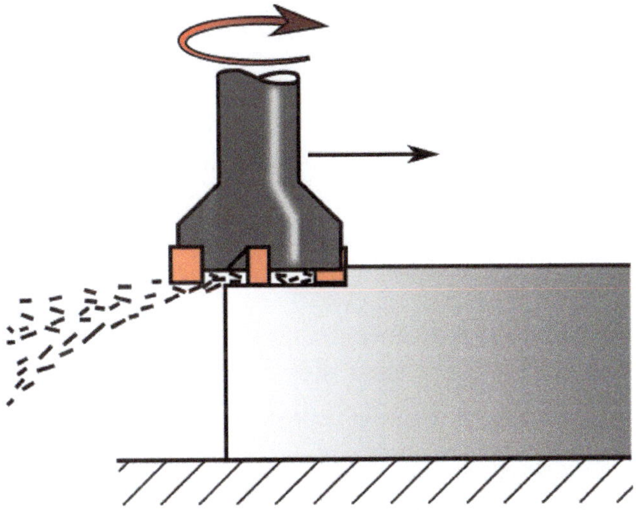

Face milling process (cutter rotation axis is vertical)

Milling is a cutting process that uses a milling cutter to remove material from the surface of a workpiece. The milling cutter is a rotary cutting tool, often with multiple cutting points. As opposed to drilling, where the tool is advanced along its rotation axis, the cutter in milling is usually moved perpendicular to its axis so that cutting occurs on the circumference of the cutter. As the milling cutter enters the workpiece, the cutting edges (flutes or teeth) of the tool repeatedly cut into and exit from the material, shaving off chips (swarf) from the workpiece with each pass. The cutting action is shear deformation; material is pushed off the workpiece in tiny clumps that hang together to a greater or lesser extent (depending on the material) to form chips. This makes metal cutting somewhat different (in its mechanics) from slicing softer materials with a blade.

The milling process removes material by performing many separate, small cuts. This is accomplished by using a cutter with many teeth, spinning the cutter at high speed, or advancing the material through the cutter slowly; most often it is some combination of these three approaches. The speeds and feeds used are varied to suit a combination of variables. The speed at which the piece advances through the cutter is called feed rate, or just feed; it is most often measured in length of material per full revolution of the cutter.

There are two major classes of milling process:

- In face milling, the cutting action occurs primarily at the end corners of the milling cutter. Face milling is used to cut flat surfaces (faces) into the work-piece, or to cut flat-bottomed cavities.

- In peripheral milling, the cutting action occurs primarily along the circumference of the cutter, so that the cross section of the milled surface ends up receiving the shape of the cutter. In this case the blades of the cutter can be seen as scooping out material from the work piece. Peripheral milling is well suited to the cutting of deep slots, threads, and gear teeth.

Milling Cutters

Many different types of cutting tools are used in the milling process. Milling cutters such as endmills may have cutting surfaces across their entire end surface, so that they can be drilled into the workpiece (plunging). Milling cutters may also have extended cutting surfaces on their sides to allow for peripheral milling. Tools optimized for face milling tend to have only small cutters at their end corners.

The cutting surfaces of a milling cutter are generally made of a hard and temperature-resistant material, so that they wear slowly. A low cost cutter may have surfaces made of high speed steel. More expensive but slower-wearing materials include cemented carbide. Thin film coatings may be applied to decrease friction or further increase hardness.

They are cutting tools typically used in milling machines or machining centres to perform milling operations (and occasionally in other machine tools). They remove material by their movement within the machine (e.g., a ball nose mill) or directly from the cutter's shape (e.g., a form tool such as a hobbing cutter).

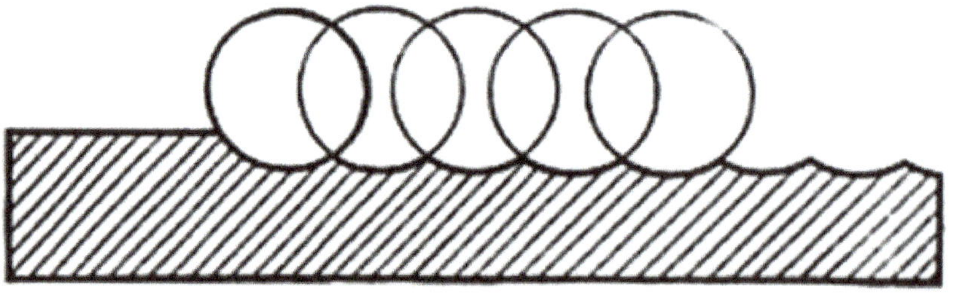

A diagram of revolution ridges on a surface milled by the side of the cutter, showing the position of the cutter for each cutting pass and how it corresponds with the ridges (cutter rotation axis is perpendicular to image plane)

As material passes through the cutting area of a milling machine, the blades of the cutter take swarfs of material at regular intervals. Surfaces cut by the side of the cutter (as in peripheral milling) therefore always contain regular ridges. The distance between ridges and the height of the ridges depend on the feed rate, number of cutting surfaces, the cutter diameter. With a narrow cutter and rapid feed rate, these revolution ridges can be significant variations in the surface finish.

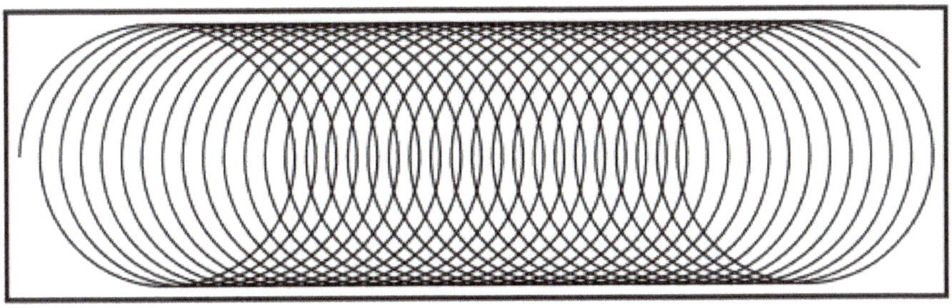

Trochoidal marks, characteristic of face milling.

The face milling process can in principle produce very flat surfaces. However, in practice the result always shows visible trochoidal marks following the motion of points on the cutter's end face. These revolution marks give the characteristic finish of a face milled surface. Revolution marks can have significant roughness depending on factors such as flatness of the cutter's end face and the degree of perpendicularity between the cutter's rotation axis and feed direction. Often a final pass with a slow feed rate is used to improve the surface finish after the bulk of the material has been removed.. In a precise face milling operation, the revolution marks will only be microscopic scratches due to imperfections in the cutting edge.

Gang Milling

Gang milling refers to the use of two or more milling cutters mounted on the same arbor (that is, ganged) in a horizontal-milling setup. All of the cutters may perform the same type of operation, or each cutter may perform a different type of operation. For example, if several workpieces need a slot, a flat surface, and an angular groove, a good method to cut these (within a non-CNC context) would be gang milling. All the completed workpieces would be the same, and milling time per piece would be minimized.

Heavy gang milling of milling machine tables

Gang milling was especially important before the CNC era, because for duplicate part production, it was a substantial efficiency improvement over manual-milling one feature at an operation, then changing machines (or changing setup of the same machine) to cut the next op. Today, CNC mills with automatic tool change and 4- or 5-axis control obviate gang-milling practice to a large extent.

Equipment

Milling is performed with a milling cutter in various forms, held in a collett or similar which, in turn, is held in the spindle of a milling machine.

Types and Nomenclature

Mill orientation is the primary classification for milling machines. The two basic configurations are vertical and horizontal. However, there are alternative classifications according to method of control, size, purpose and power source.

Mill Orientation

Vertical Mill

In the vertical mill the spindle axis is vertically oriented. Milling cutters are held in the spindle and rotate on its axis. The spindle can generally be extended (or the table can be raised/lowered, giving the same effect), allowing plunge cuts and drilling. There are two subcategories of vertical mills: the bed mill and the turret mill.

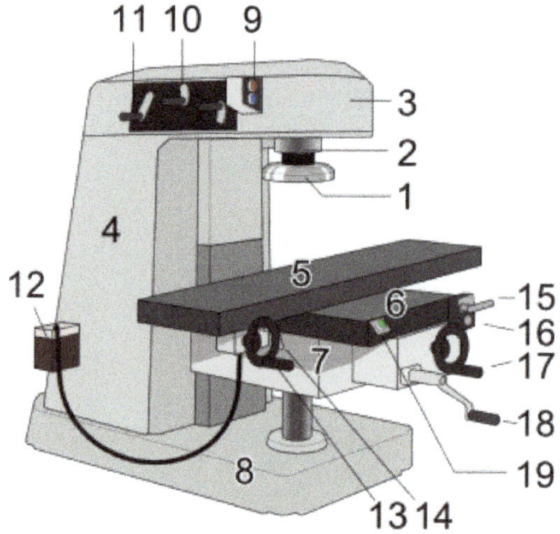

Vertical milling machine. 1: milling cutter 2: spindle 3: top slide or overarm 4: column 5: table 6: Y-axis slide 7: knee 8: base

- A turret mill has a stationary spindle and the table is moved both perpendicular and parallel to the spindle axis to accomplish cutting. The most common example of this type is the Bridgeport, described below. Turret mills often have a quill which allows the milling cutter to be raised and lowered in a manner similar to a drill press. This type of machine provides two methods of cutting in the vertical (Z) direction: by raising or lowering the quill, and by moving the knee.

- In the bed mill, however, the table moves only perpendicular to the spindle's axis, while the spindle itself moves parallel to its own axis.

Turret mills are generally considered by some to be more versatile of the two designs. However, turret mills are only practical as long as the machine remains relatively small. As machine size increases, moving the knee up and down requires considerable effort and it also becomes difficult to reach the quill feed handle (if equipped). Therefore, larger milling machines are usually of the bed type.

A third type also exists, a lighter machine, called a mill-drill, which is a close relative of the vertical mill and quite popular with hobbyists. A mill-drill is similar in basic configuration to a small drill press, but equipped with an X-Y table. They also typically use more powerful motors than a comparably sized drill press, with potentiometer-controlled speed and generally have more heavy-duty spindle bearings than a drill press to deal with the lateral loading on the spindle that is created by a milling operation. A mill drill also typically raises and lowers the entire head, including motor, often on a dovetailed vertical, where a drill press motor remains stationary, while the arbor raises and lowers within a driving collar. Other differences that separate a mill-drill from a drill press may be a fine tuning adjustment for the Z-axis, a more precise depth stop, the capability to lock the X, Y or Z axis, and often a system of tilting the head or the entire vertical column and powerhead assembly to allow angled cutting. Aside from size and precision, the principal difference between these hobby-type machines and larger true vertical mills is that the X-Y table is at a fixed elevation; the Z-axis is controlled in basically the same fashion as drill press, where a larger vertical or knee mill has a vertically fixed milling head, and changes the X-Y table elevation. As well, a mill-drill often uses a standard drill press-type Jacob's chuck, rather than an internally tapered arbor that accepts collets. These are frequently of lower quality than other types of machines, but still fill the hobby role well because they tend to be benchtop machines with small footprints and modest price tags.

Horizontal Mill

A horizontal mill has the same sort but the cutters are mounted on a horizontal arbor across the table. Many horizontal mills also feature a built-in rotary table that allows milling at various angles; this feature is called a *universal table*. While endmills and the other types of tools available to a vertical mill may be used in a horizontal mill, their real advantage lies in arbor-mounted cutters, called side and face mills, which have a cross section rather like a circular saw, but are generally wider and smaller in diameter. Because the cutters have good support from the arbor and have a larger cross-sectional area than an end mill, quite heavy cuts can be taken enabling rapid material removal rates. These are used to mill grooves and slots. Plain mills are used to shape flat surfaces. Several cutters may be ganged together on the arbor to mill a complex shape of slots and planes. Special cutters can also cut grooves, bevels, radii, or indeed any section desired. These specialty cutters tend to be expensive. Simplex mills have one spindle, and duplex mills have two. It is also easier to cut gears on a horizontal mill. Some horizontal milling machines are equipped with a power-take-off provision on the table. This allows the table feed to be synchronized to a rotary fixture, enabling the milling of spiral features such as hypoid gears.

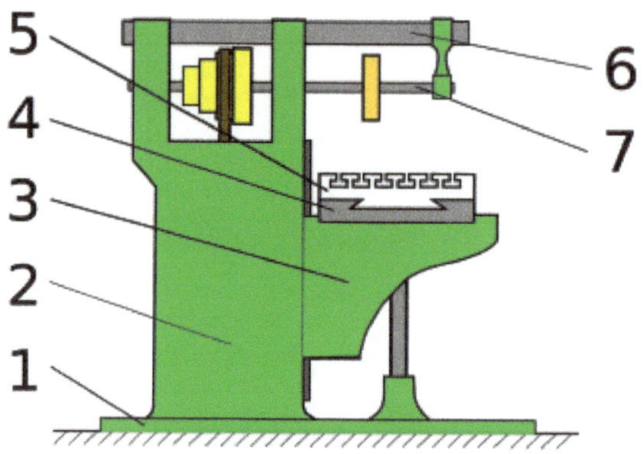

Horizontal milling machine. 1: base 2: column 3: knee 4 & 5: table (x-axis slide is integral) 6: overarm 7: arbor (attached to spindle)

Comparative Merits

The choice between vertical and horizontal spindle orientation in milling machine design usually hinges on the shape and size of a workpiece and the number of sides of the workpiece that require machining. Work in which the spindle's axial movement is normal to one plane, with an endmill as the cutter, lends itself to a vertical mill, where the operator can stand before the machine and have easy access to the cutting action by looking down upon it. Thus vertical mills are most favored for diesinking work (machining a mould into a block of metal). Heavier and longer workpieces lend themselves to placement on the table of a horizontal mill.

Prior to numerical control, horizontal milling machines evolved first, because they evolved by putting milling tables under lathe-like headstocks. Vertical mills appeared in subsequent decades, and accessories in the form of add-on heads to change horizontal mills to vertical mills (and later vice versa) have been commonly used. Even in the CNC era, a heavy workpiece needing machining on multiple sides lends itself to a horizontal machining center, while diesinking lends itself to a vertical one.

Alternative Classifications

In addition to horizontal versus vertical, other distinctions are also important:

Criterion	Example classification scheme	Comments
Spindle axis orientation	Vertical versus horizontal; Turret versus non-turret	Among vertical mills, "Bridgeport-style" is a whole class of mills inspired by the Bridgeport original, rather like the IBM PC spawned the industry of IBM-compatible PCs by other brands
Control	Manual; Mechanically automated via cams; Digitally automated via NC/CNC	In the CNC era, a very basic distinction is manual versus CNC. Among manual machines, a worthwhile distinction is non-DRO-equipped versus DRO-equipped
Control (specifically among CNC machines)	Number of axes (e.g., 3-axis, 4-axis, or more)	Within this scheme, also: • Pallet-changing versus non-pallet-changing • Full-auto tool-changing versus semi-auto or manual tool-changing
Purpose	General-purpose versus special-purpose or single-purpose	
Purpose	Toolroom machine versus production machine	Overlaps with above
Purpose	"Plain" versus "universal"	A distinction whose meaning evolved over decades as technology progressed, and overlaps with other purpose classifications above. Not relevant to today's CNC mills. Regarding manual mills, the common theme is that "plain" mills were production machines with fewer axes than "universal" mills; for example, whereas a plain mill had no indexing head and a non-rotating table, a universal mill would have those. Thus it was suited to universal service, that is, a wider range of possible toolpaths. Machine tool builders no longer use the "plain"-versus-"universal" labeling.
Size	Micro, mini, benchtop, standing on floor, large, very large, gigantic	
Power source	Line-shaft-drive versus individual electric motor drive	Most line-shaft-drive machines, ubiquitous circa 1880–1930, have been scrapped by now
	Hand-crank-power versus electric	Hand-cranked not used in industry but suitable for hobbyist micromills

Variants

- Bed mill This refers to any milling machine where the spindle is on a pendant that moves up and down to move the cutter into the work, while the table sits on a stout bed that rests on the floor. These are generally more rigid than a knee mill. Gantry mills can be included in this bed mill category.

- Box mill or column mill Very basic hobbyist bench-mounted milling machines that feature a head riding up and down on a column or box way.
- C-frame mill These are larger, industrial production mills. They feature a knee and fixed spindle head that is only mobile vertically. They are typically much more powerful than a turret mill, featuring a separate hydraulic motor for integral hydraulic power feeds in all directions, and a twenty to fifty horsepower motor. Backlash eliminators are almost always standard equipment. They use large NMTB 40 or 50 tooling. The tables on C-frame mills are usually 18" by 68" or larger, to allow multiple parts to be machined at the same time.
- Floor mill These have a row of rotary tables, and a horizontal pendant spindle mounted on a set of tracks that runs parallel to the table row. These mills have predominantly been converted to CNC, but some can still be found (if one can even find a used machine available) under manual control. The spindle carriage moves to each individual table, performs the machining operations, and moves to the next table while the previous table is being set up for the next operation. Unlike other mills, floor mills have movable floor units. A crane drops massive rotary tables, X-Y tables, etc., into position for machining, allowing large and complex custom milling operations.

A Sieg X2 miniature hobbyist mill plainly showing the basic parts of a mill.

- Gantry mill The milling head rides over two rails (often steel shafts) which lie at each side of the work surface.
- Horizontal boring mill Large, accurate bed horizontal mills that incorporate many features from various machine tools. They are predominantly used to create large manufacturing jigs, or to modify large, high precision parts. They have a spindle stroke of several (usually between four and six) feet, and many are equipped with a tailstock to perform very long boring operations without losing

accuracy as the bore increases in depth. A typical bed has X and Y travel, and is between three and four feet square with a rotary table or a larger rectangle without a table. The pendant usually provides between four and eight feet of vertical movement. Some mills have a large (30" or more) integral facing head. Right angle rotary tables and vertical milling attachments are available for further flexibility.

- Jig borer Vertical mills that are built to bore holes, and very light slot or face milling. They are typically bed mills with a long spindle throw. The beds are more accurate, and the handwheels are graduated down to .0001" for precise hole placement.
- Knee mill or knee-and-column mill refers to any milling machine whose x-y table rides up and down the column on a vertically adjustable knee. This includes Bridgeports.
- Planer-style mill Large mills built in the same configuration as planers except with a milling spindle instead of a planing head. This term is growing dated as planers themselves are largely a thing of the past.
- Ram-type mill This can refer to any mill that has a cutting head mounted on a sliding ram. The spindle can be oriented either vertically or horizontally. In practice most mills with rams also involve swiveling ability, whether or not it is called "turret" mounting. The Bridgeport configuration can be classified as a vertical-head ram-type mill. Van Norman specialized in ram-type mills through most of the 20th century. Since the wide dissemination of CNC machines, ram-type mills are still made in the Bridgeport configuration (with either manual or CNC control), but the less common variations (such as were built by Van Norman, Index, and others) have died out, their work being done now by either Bridgeport-form mills or machining centers.
- Turret mill More commonly referred to as Bridgeport-type milling machines. The spindle can be aligned in many different positions for a very versatile, if somewhat less rigid machine.

Alternative Terminology

A milling machine is often called a mill by machinists. The archaic term miller was commonly used in the 19th and early 20th centuries.

Since the 1960s there has developed an overlap of usage between the terms milling machine and machining center. NC/CNC machining centers evolved from milling machines, which is why the terminology evolved gradually with considerable overlap that still persists. The distinction, when one is made, is that a machining center is a mill with features that pre-CNC mills never had, especially an automatic tool changer (ATC) that includes a tool magazine (carousel), and sometimes an automatic pallet changer (APC). In typical usage, all machining centers are mills, but not all mills are machining centers; only mills with ATCs are machining centers.

Computer Numerical Control

Thin wall milling of aluminum using a water based cutting fluid on the milling cutter

Most CNC milling machines (also called *machining centers*) are computer controlled vertical mills with the ability to move the spindle vertically along the Z-axis. This extra degree of freedom permits their use in diesinking, engraving applications, and 2.5D surfaces such as relief sculptures. When combined with the use of conical tools or a ball nose cutter, it also significantly improves milling precision without impacting speed, providing a cost-efficient alternative to most flat-surface hand-engraving work.

Five-axis machining center with rotating table and computer interface

CNC machines can exist in virtually any of the forms of manual machinery, like horizontal mills. The most advanced CNC milling-machines, the multiaxis machine, add two more axes in addition to the three normal axes (XYZ). Horizontal milling machines also have a C or Q axis, allowing the horizontally mounted workpiece to be rotated, essentially allowing asymmetric and eccentric turning. The fifth axis (B axis) controls the tilt of the tool itself. When all of these axes are used in conjunction with each other, extremely complicated geometries, even organic geometries such as a human head can be made with relative ease with these machines. But the skill to program such geometries is beyond that of most operators. Therefore, 5-axis milling machines are practically always programmed with CAM.

The operating system of such machines is a closed loop system and functions on feedback. These machines have developed from the basic NC (NUMERIC CONTROL) machines. A computerized form of NC machines is known as CNC machines. A set of instructions (called a program) is used to guide the machine for desired operations. Some very commonly used codes, which are used in the program are:

G00 – rapid traverse

G01 – linear interpolation of tool.

G21 – dimensions in metric units.

M03/M04 – spindle start (clockwise/counter clockwise).

T01 M06 – automatic tool change to tool 1

M30 – program end.

Various other codes are also used. A CNC machine is operated by a single operator called a programmer. This machine is capable of performing various operations automatically and economically.

With the declining price of computers and open source CNC software, the entry price of CNC machines has plummeted.

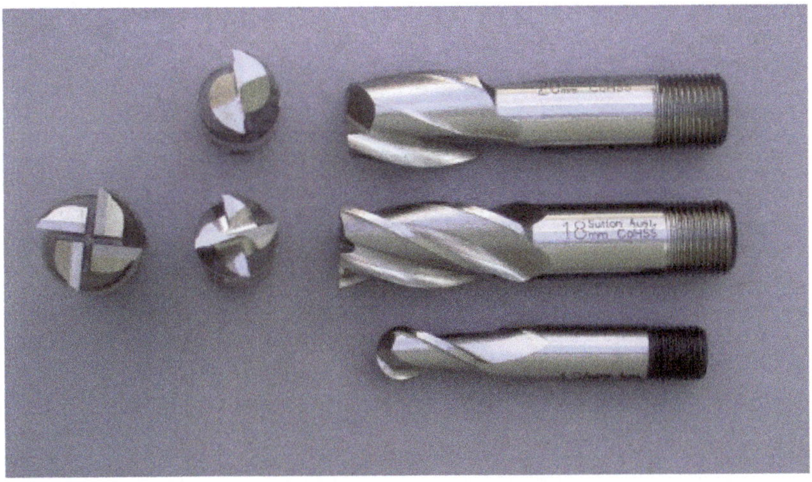

High speed steel with cobalt endmills used for cutting operations in a milling machine.

Tooling

The accessories and cutting tools used on machine tools (including milling machines) are referred to in aggregate by the mass noun "tooling". There is a high degree of standardization of the tooling used with CNC milling machines, and a lesser degree with manual milling machines. To ease up the organization of the tooling in CNC production many companies use a tool management solution.

Milling cutters for specific applications are held in various tooling configurations.

CNC milling machines nearly always use SK (or ISO), CAT, BT or HSK tooling. SK tooling is the most common in Europe, while CAT tooling, sometimes called V-Flange Tooling, is the oldest and probably most common type in the USA. CAT tooling was invented by Caterpillar Inc. of Peoria, Illinois, in order to standardize the tooling used on their machinery. CAT tooling comes in a range of sizes designated as CAT-30, CAT-40, CAT-50, etc. The number refers to the Association for Manufacturing Technology (formerly the National Machine Tool Builders Association (NMTB)) Taper size of the tool.

A CAT-40 toolholder

An improvement on CAT Tooling is BT Tooling, which looks similar and can easily be confused with CAT tooling. Like CAT Tooling, BT Tooling comes in a range of sizes and uses the same NMTB body taper. However, BT tooling is symmetrical about the spindle axis, which CAT tooling is not. This gives BT tooling greater stability and balance at high speeds. One other subtle difference between these two toolholders is the thread used to hold the pull stud. CAT Tooling is all Imperial thread and BT Tooling is all Metric thread. Note that this affects the pull stud only, it does not affect the tool that they can hold, both types of tooling are sold to accept both Imperial and metric sized tools.

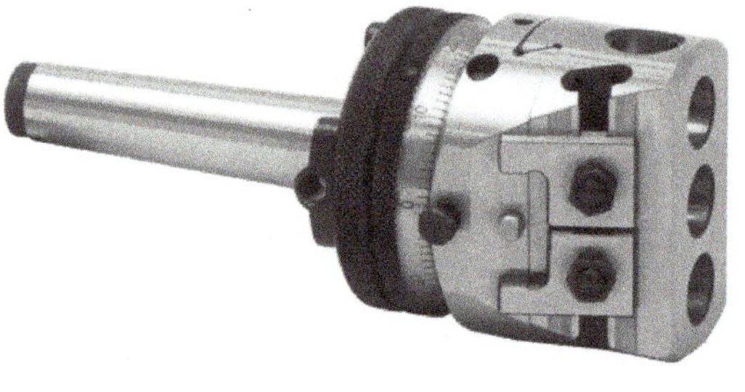

A boring head on a Morse taper shank

SK and HSK tooling, sometimes called "Hollow Shank Tooling", is much more common in Europe where it was invented than it is in the United States. It is claimed that HSK tooling is even better than BT Tooling at high speeds. The holding mechanism for HSK tooling is placed within the (hollow) body of the tool and, as spindle speed increases, it expands, gripping the tool more tightly with increasing spindle speed. There is no pull stud with this type of tooling.

For manual milling machines, there is less standardization, because a greater plurality of formerly competing standards exist. Newer and larger manual machines usually use NMTB tooling. This tooling is somewhat similar to CAT tooling but requires a drawbar within the milling machine. Furthermore, there are a number of variations with NMTB tooling that make interchangeability troublesome. The older a machine, the greater the plurality of standards that may apply (e.g., Morse, Jarno, Brown & Sharpe, Van Norman, and other less common builder-specific tapers). However, two standards that have seen especially wide usage are the Morse #2 and the R8, whose prevalence was driven by the popularity of the mills built by Bridgeport Machines of Bridgeport, Connecticut. These mills so dominated the market for such a long time that "Bridgeport" is virtually synonymous with "manual milling machine". Most of the machines that Bridgeport made between 1938 and 1965 used a Morse taper #2, and from about 1965 onward most used an R8 taper.

Accessories

- Arbor support
- Stop block

History

1810s–1830s

Milling machines evolved from the practice of rotary filing—that is, running a circular cutter with file-like teeth in the headstock of a lathe. Rotary filing and, later, true milling were developed to reduce time and effort spent hand-filing. The full story of milling machine development may never be known, because much early development took place in individual shops where few records were kept for posterity. However, the broad outlines are known, as summarized below. From a history-of-technology viewpoint, it is clear that the naming of this new type of machining with the term "milling" was an extension from that word's earlier senses of processing materials by abrading them in some way (cutting, grinding, crushing, etc.).

Rotary filing long predated milling. A rotary file by Jacques de Vaucanson, circa 1760, is well known. It is clear that milling machines as a distinct class of machine tool (separate from lathes running rotary files) first appeared between 1814 and 1818. The centers of earliest development of true milling machines were two federal armories of the U.S.

(Springfield and Harpers Ferry) together with the various private armories and inside contractors that shared turnover of skilled workmen with them.

This milling machine was long credited to Eli Whitney and dated to circa 1818. From the 1910s through the 1940s, this version of its provenance was widely published. In the 1950s and 1960s, various historians of technology mostly discredited the view of this machine as the first miller and possibly even of Whitney as its builder. Nonetheless, it is still an important early milling machine, regardless of its exact provenance.

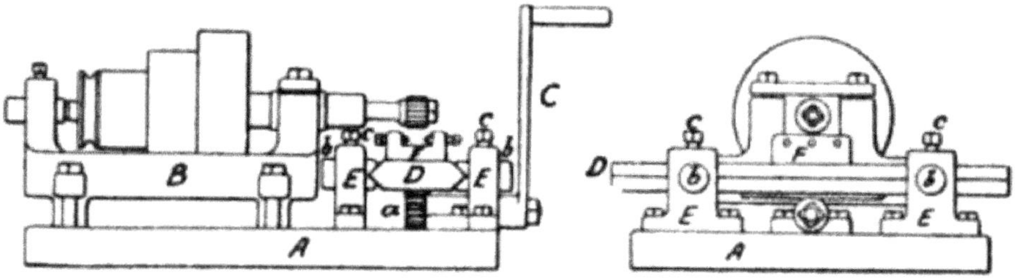

The Middletown milling machine of circa 1818, associated with Robert Johnson and Simeon North.

Between 1912 and 1916, Joseph W. Roe, a respected founding father of machine tool historians, credited Eli Whitney (one of the private arms makers mentioned above) with producing the first true milling machine. By 1918, he considered it "Probably the first milling machine ever built—certainly the oldest now in existence [...]." However, subsequent scholars, including Robert S. Woodbury and others, have improved upon Roe's early version of the history and suggest that just as much credit—in fact, probably more—belongs to various other inventors, including Robert Johnson of Middletown, Connecticut; Captain John H. Hall of the Harpers Ferry armory; Simeon North of the Staddle Hill factory in Middletown; Roswell Lee of the Springfield armory; and Thomas Blanchard. (Several of the men mentioned above are sometimes described on the internet as "the inventor of the first milling machine" or "the inventor of interchangeable parts". Such claims are oversimplified, as these technologies evolved over time among many people.)

Peter Baida, citing Edward A. Battison's article "Eli Whitney and the Milling Machine," which was published in the *Smithsonian Journal of History* in 1966, exemplifies the dispelling of the "Great Man" image of Whitney by historians of technology working in the 1950s and 1960s. He quotes Battison as concluding that "There is no evidence that Whitney developed or used a true milling machine." Baida says, "The so-called Whitney machine of 1818 seems actually to have been made after Whitney's death in 1825." Baida cites Battison's suggestion that the first true milling machine was made not by Whitney, but by Robert Johnson of Middletown.

The late teens of the 19th century were a pivotal time in the history of machine tools, as the period of 1814 to 1818 is also the period during which several contemporary pioneers (Fox, Murray, and Roberts) were developing the planer, and as with the milling machine, the work being done in various shops was undocumented for various reasons (partially because of proprietary secrecy, and also simply because no one was taking down records for posterity).

James Nasmyth built a milling machine very advanced for its time between 1829 and 1831. It was tooled to mill the six sides of a hex nut that was mounted in a six-way indexing fixture.

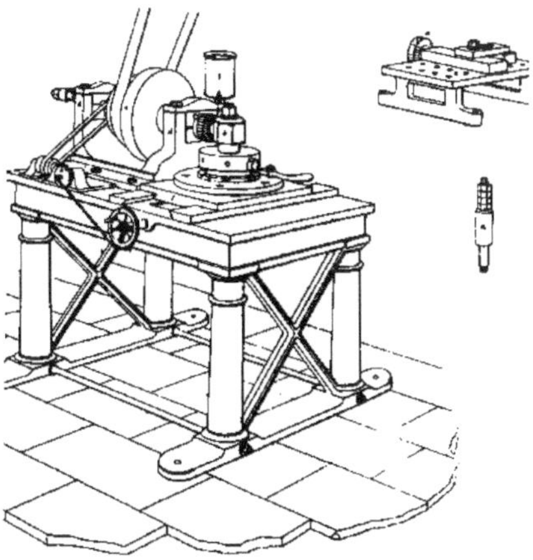

The milling machine built by James Nasmyth between 1829 and 1831 for milling the six sides of a hex nut using an indexing fixture.

A milling machine built and used in the shop of Gay & Silver (aka Gay, Silver, & Co) in the 1830s was influential because it employed a better method of vertical positioning than earlier machines. For example, Whitney's machine (the one that Roe considered the very first) and others did not make provision for vertical travel of the knee. Evidently, the workflow assumption behind this was that the machine would be set up with shims, vise, etc. for a certain part design, and successive parts did not require vertical adjustment (or at most would need only shimming). This indicates that early thinking about milling machines was as production machines, not toolroom machines.

In these early years, milling was often viewed as only a roughing operation to be followed by finishing with a hand file. The idea of *reducing* hand filing was more important than *replacing* it.

1840s–1860

Some of the key men in milling machine development during this era included Frederick W. Howe, Francis A. Pratt, Elisha K. Root, and others. (These same men during the same era were also busy developing the state of the art in turret lathes. Howe's experience at Gay & Silver in the 1840s acquainted him with early versions of both machine tools. His machine tool designs were later built at Robbins & Lawrence, the Providence Tool Company, and Brown & Sharpe.) The most successful milling machine design to emerge during this era was the Lincoln miller, which rather than being a specific make and model of machine tool is truly a family of tools built by various companies on a common configuration over several decades. It took its name from the first company to put one on the market, George S. Lincoln & Company (formerly the Phoenix Iron Works), whose first one was built in 1855 for the Colt armory.

A typical Lincoln miller. The configuration was established in the 1850s. (This example was built by Pratt & Whitney, probably 1870s or 1880s.)

During this era there was a continued blind spot in milling machine design, as various designers failed to develop a truly simple and effective means of providing slide travel in all three of the archetypal milling axes (X, Y, and Z—or as they were known in the past, longitudinal, traverse, and vertical). Vertical positioning ideas were either absent or underdeveloped. The Lincoln miller's spindle could be raised and lowered, but the original idea behind its positioning was to be set up in position and then run, as opposed to being moved frequently while running. Like a turret lathe, it was a repetitive-production machine, with each skilled setup followed by extensive fairly low skill operation.

1860s

In 1861, Frederick W. Howe, while working for the Providence Tool Company, asked Joseph R. Brown of Brown & Sharpe for a solution to the problem of milling spirals, such as the flutes of twist drills. These were usually filed by hand at the time. (Helical planing existed but was by no means common.) Brown designed a "universal milling machine" that, starting from its first sale in March 1862, was wildly successful. It solved the problem of 3-axis travel (i.e., the axes that we now call XYZ) much more elegantly than had been done in the past, and it allowed for the milling of spirals using an indexing head fed in coordination with the table feed. The term "universal" was applied to it because it was ready for any kind of work, including toolroom work, and was not as limited in application as previous designs. (Howe had designed a "universal miller" in 1852, but Brown's of 1861 is the one considered a groundbreaking success.)

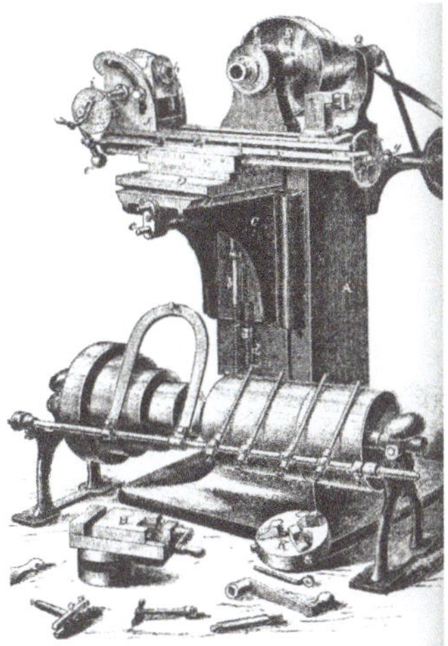

Brown & Sharpe's groundbreaking universal milling machine, 1861

Brown also developed and patented (1864) the design of formed milling cutters in which successive sharpenings of the teeth do not disturb the geometry of the form.

The advances of the 1860s opened the floodgates and ushered in modern milling practice.

1870s to World War I

A typical universal milling machine of the early 20th century. Suitable for toolroom, jobbing, or production use.

In these decades, Brown & Sharpe and the Cincinnati Milling Machine Company dominated the milling machine field. However, hundreds of other firms also built milling machines at the time, and many were significant in various ways. Besides a wide variety of specialized production machines, the archetypal multipurpose milling machine of the late 19th and early 20th centuries was a heavy knee-and-column horizontal-spindle design with power table feeds, indexing head, and a stout overarm to support the arbor. The evolution of machine design was driven not only by inventive spirit but also by the constant evolution of milling cutters that saw milestone after milestone from 1860 through World War I.

World War I and Interwar Period

Around the end of World War I, machine tool control advanced in various ways that laid the groundwork for later CNC technology. The jig borer popularized the ideas of coordinate dimensioning (dimensioning of all locations on the part from a single reference point); working routinely in "tenths" (ten-thousandths of an inch, 0.0001") as an everyday machine capability; and using the control to go straight from drawing to part, circumventing jig-making. In 1920 the new tracer design of J.C. Shaw was applied to Keller tracer milling machines for die-sinking via the three-dimensional copying of a template. This made diesinking faster and easier just as dies were in higher demand than ever before, and was very helpful for large steel dies such as those used to stamp sheets in automobile manufacturing. Such machines translated the tracer movements to input for servos that worked the machine leadscrews or hydraulics. They also spurred the development of antibacklash leadscrew nuts. All of the above concepts were new in the 1920s but became routine in the NC/CNC era. By the 1930s, incredibly large and advanced milling machines existed, such as the Cincinnati Hydro-Tel, that presaged today's CNC mills in every respect except for CNC control itself.

Bridgeport Milling Machine

In 1936, Rudolph Bannow (1897–1962) conceived of a major improvement to the milling machine. His company commenced manufacturing a new knee-and-column vertical mill in 1938. This was the Bridgeport milling machine, often called a ram-type or turret-type mill because its head has sliding-ram and rotating-turret mounting. The machine became so popular that many other manufacturers created copies and variants. Furthermore, its name came to connote any such variant. The Bridgeport offered enduring advantages over previous models. It was small enough, light enough, and affordable enough to be a practical acquisition for even the smallest machine shop businesses, yet it was also smartly designed, versatile, well-built, and rigid. Its various directions of sliding and pivoting movement allowed the head to approach the work from any angle. The Bridgeport's design became the dominant form for manual milling machines used by several generations of small- and medium-enterprise machinists. By the 1980s an estimated quarter-million Bridgeport milling machines had been built, and they (and their clones) are still being produced today.

1940s–1970s

By 1940, automation via cams, such as in screw machines and automatic chuckers, had already been very well developed for decades. Beginning in the 1930s, ideas involving servomechanisms had been in the air, but it was especially during and immediately after World War II that they began to germinate. These were soon combined with the emerging technology of digital computers. This technological development milieu, spanning from the immediate pre–World War II period into the 1950s, was powered by the military capital expenditures that pursued contemporary advancements in the directing of gun and rocket artillery and in missile guidance—other applications in which humans wished to control the kinematics/dynamics of large machines quickly, precisely, and automatically. Sufficient R&D spending probably would not have happened within the machine tool industry alone; but it was for the latter applications that the will and ability to spend was available. Once the development was underway, it was eagerly applied to machine tool control in one of the many post-WWII instances of technology transfer.

In 1952, numerical control reached the developmental stage of laboratory reality. The first NC machine tool was a Cincinnati Hydrotel milling machine retrofitted with a scratch-built NC control unit. It was reported in *Scientific American*, just as another groundbreaking milling machine, the Brown & Sharpe universal, had been in 1862.

During the 1950s, numerical control moved slowly from the laboratory into commercial service. For its first decade, it had rather limited impact outside of aerospace work. But during the 1960s and 1970s, NC evolved into CNC, data storage and input media evolved, computer processing power and memory capacity steadily increased, and NC and CNC machine tools gradually disseminated from an environment of huge corporations and mainly aerospace work to the level of medium-sized corporations and a wide variety of products. NC and CNC's drastic advancement of machine tool control deeply transformed the culture of manufacturing. The details have evolved immensely with every passing decade.

1980s–present

Computers and CNC machine tools continue to develop rapidly. The personal computer revolution has a great impact on this development. By the late 1980s small machine shops had desktop computers and CNC machine tools. Soon after, hobbyists, artists, and designers began obtaining CNC mills and lathes. Manufacturers have started producing economically priced CNCs machines small enough to sit on a desktop which can cut at high resolution materials softer than stainless steel. They can be used to make anything from jewelry to printed circuit boards to gun parts, even fine art.

Milling Standards

National and international standards are used to standardize the definitions, environmental requirements, and test methods used for milling. Selection of the standard to be used is an agreement between the supplier and the user and has some significance in the design of the mill. In the United States, ASME has developed the B5.45-1972 Standard entitled "Milling Machines", B94.19-1997, Milling Cutters and End Mills.

Machine Tool

A machine tool is a machine for shaping or machining metal or other rigid materials, usually by cutting, boring, grinding, shearing, or other forms of deformation. Machine tools employ some sort of tool that does the cutting or shaping. All machine tools have some means of constraining the workpiece and provide a guided movement of the parts of the machine. Thus the relative movement between the workpiece and the cutting tool (which is called the toolpath) is controlled or constrained by the machine to at least some extent, rather than being entirely "offhand" or "freehand".

A lathe is an example of a machine tool

The precise definition of the term *machine tool* varies among users, as discussed below. While all machine tools are "machines that help people to make things", not all factory machines are machine tools.

Today machine tools are typically powered other than by human muscle (e.g., electrically, hydraulically, or via line shaft), used to make manufactured parts (components) in various ways that include cutting or certain other kinds of deformation.

With their inherent precision, machine tools enabled the economical production of interchangeable parts.

Nomenclature and Key Concepts, Interrelated

Many historians of technology consider that true machine tools were born when the toolpath first became guided by the machine itself in some way, at least to some extent, so that direct, freehand human guidance of the toolpath (with hands, feet, or mouth) was no longer the only guidance used in the cutting or forming process. In this view of the definition, the term, arising at a time when all tools up till then had been hand tools, simply provided a label for "tools that were machines instead of hand tools". Early lathes, those prior to the late medieval period, and modern woodworking lathes and potter's wheels may or may not fall under this definition, depending on how one views the headstock spindle itself; but the earliest historical records of a lathe with direct mechanical control *of the cutting tool's path* are of a screw-cutting lathe dating to about 1483. This lathe "produced screw threads out of wood and employed a true compound slide rest".

The mechanical toolpath guidance grew out of various root concepts:

- First is the spindle concept itself, which constraints workpiece or tool movement to rotation around a fixed axis. This ancient concept predates machine tools per se; the earliest lathes and potter's wheels incorporated it for the workpiece, but the movement of the tool itself on these machines was entirely freehand.
- The machine slide, which has many forms, such as dovetail ways, box ways, or cylindrical column ways. Machine slides constrain tool or workpiece movement linearly. If a stop is added, the *length* of the line can also be accurately controlled. (Machine slides are essentially a subset of linear bearings, although the language used to classify these various machine elements includes connotative boundaries; some users in some contexts would contradistinguish elements in ways that others might not.)
- Tracing, which involves following the contours of a model or template and transferring the resulting motion to the toolpath.
- Cam operation, which is related in principle to tracing but can be a step or two removed from the traced element's matching the reproduced element's final shape. For example, several cams, no one of which directly matches the desired output shape, can actuate a complex toolpath by creating component vectors that add up to a net toolpath.

Abstractly programmable toolpath guidance began with mechanical solutions, such as in musical box cams and Jacquard looms. The convergence of programmable mechanical control with machine tool toolpath control was delayed many decades, in part because the programmable control methods of musical boxes and looms lacked the rigidity for machine tool toolpaths. Later, electromechanical solutions (such as servos) and soon electronic solutions (including computers) were added, leading to numerical control and computer numerical control.

When considering the difference between freehand toolpaths and machine-constrained toolpaths, the concepts of accuracy and precision, efficiency, and productivity become important in understanding *why* the machine-constrained option adds value. After all, humans are generally quite talented in their freehand movements; the drawings, paintings, and sculptures of artists such as Michelangelo or Leonardo da Vinci, and of countless other talented people, show that human freehand toolpath has great potential. The value that machine tools added to these human talents is in the areas of rigidity (constraining the toolpath despite thousands of newtons (pounds) of force fighting against the constraint), accuracy and precision, efficiency, and productivity. With a machine tool, toolpaths that no human muscle could constrain can be constrained; and toolpaths that are technically possible with freehand methods, but would require tremendous time and skill to execute, can instead be executed quickly and easily, even by people with little freehand talent (because the machine takes care of it). The latter aspect of machine tools is often referred to by historians of technology as "building the skill into the tool", in contrast to the toolpath-constraining skill being in the *person* who wields the tool. As an example, it is *physically possible* to make interchangeable screws, bolts, and nuts entirely with freehand toolpaths. But it is *economically practical* to make them only with machine tools.

In the 1930s, the U.S. National Bureau of Economic Research (NBER) referenced the definition of a machine tool as "any machine operating by other than hand power which employs a tool to work on metal".

The narrowest colloquial sense of the term reserves it only for machines that perform metal cutting—in other words, the many kinds of [conventional] machining and grinding. These processes are a type of deformation that produces swarf. However, economists use a slightly broader sense that also includes metal deformation of other types that squeeze the metal into shape without cutting off swarf, such as rolling, stamping with dies, shearing, swaging, riveting, and others. Thus presses are usually included in the economic definition of machine tools. For example, this is the breadth of definition used by Max Holland in his history of Burgmaster and Houdaille, which is also a history of the machine tool industry in general from the 1940s through the 1980s; he was reflecting the sense of the term used by Houdaille itself and other firms in the industry. Many reports on machine tool export and import and similar economic topics use this broader definition.

The colloquial sense implying [conventional] metal cutting is also growing obsolete because of changing technology over the decades. The many more recently developed processes labeled "machining", such as electrical discharge machining, electrochemical machining, electron beam machining, photochemical machining, and ultrasonic machining, or even plasma cutting and water jet cutting, are often performed by machines that could most logically be called machine tools. In addition, some of the newly developed additive manufacturing processes, which are not about cutting away material but rather about adding it, are done by machines that are likely to end up labeled, in some cases, as machine tools. In fact, machine tool builders are already developing machines that include both subtractive and additive manufacturing in one work envelope, and retrofits of existing machines are underway.

The natural language use of the terms varies, with subtle connotative boundaries. Many speakers resist using the term "machine tool" to refer to woodworking machinery (joiners, table saws, routing stations, and so on), but it is difficult to maintain any true logical dividing line, and therefore many speakers are fine with a broad definition. It is common to hear machinists refer to their machine tools simply as "machines". Usually the mass noun "machinery" encompasses them, but sometimes it is used to imply only those machines that are being excluded from the definition of "machine tool". This is why the machines in a food-processing plant, such as conveyors, mixers, vessels, dividers, and so on, may be labeled "machinery", while the machines in the factory's tool and die department are instead called "machine tools" in contradistinction. As for the 1930s NBER definition quoted above, one could argue that its specificity to metal is obsolete, as it is quite common today for particular lathes, milling machines, and machining centers (definitely machine tools) to work exclusively on plastic cutting jobs throughout their whole working lifespan. Thus the NBER definition above could be expanded to say "which employs a tool to work on metal *or other materials of high hardness*". And its specificity to "operating by other than hand power" is also problematic, as machine tools can be powered by people if appropriately set up, such as with a treadle (for a lathe) or a hand lever (for a shaper). Hand-powered shapers are clearly "the 'same thing' as shapers with electric motors except smaller", and it is trivial to power a micro lathe with a hand-cranked belt pulley instead of an electric motor. Thus one can question whether power source is truly a key distinguishing concept; but for economics purposes, the NBER's definition made sense, because most of the commercial value of the existence of machine tools comes about via those that are powered by electricity, hydraulics, and so on. Such are the vagaries of natural language and controlled vocabulary, both of which have their places in the business world.

History

Forerunners of machine tools included bow drills and potter's wheels, which had existed in ancient Egypt prior to 2500 BC, and lathes, known to have existed in multiple regions of Europe since at least 1000 to 500 BC. But it was not until the later Middle Ages and the Age of Enlightenment that the modern concept of a machine tool—a class of machines used as tools in the making of metal parts, and incorporating machine-guided toolpath—began to evolve. Clockmakers of the Middle Ages and renaissance men such as Leonardo da Vinci helped expand humans' technological milieu toward the preconditions for industrial machine tools. During the 18th and 19th centuries, and even in many cases in the 20th, the builders of machine tools tended to be the same people who would then use them to produce the end products (manufactured goods). However, from these roots also evolved an industry of machine tool builders as we define them today, meaning people who specialize in building machine tools for sale to others.

Historians of machine tools often focus on a handful of major industries that most spurred machine tool development. In order of historical emergence, they have been firearms (small arms and artillery); clocks; textile machinery; steam engines (stationary, marine, rail, and otherwise) (the story of how Watt's need for an accurate cylinder spurred Boulton's boring machine is discussed by Roe); sewing machines; bicycles; automobiles; and aircraft. Others could be included in this list as well, but they tend to be connected with the root causes already listed. For example, rolling-element bearings are an industry of themselves, but this industry's main drivers of development were the vehicles already listed—trains, bicycles, automobiles, and aircraft; and other industries, such as tractors, farm implements, and tanks, borrowed heavily from those same parent industries.

Machine tools filled a need created by textile machinery during the Industrial Revolution in England in the middle to late 1700s. Until that time machinery was made mostly from wood, often including gearing and shafts. The increase in mechanization required more metal parts, which were usually made of cast iron or wrought iron. Cast iron could be cast in molds for larger parts, such as engine cylinders and gears, but was difficult to work with a file and could not be hammered. Red hot wrought iron could be hammered into shapes. Room temperature wrought iron was worked with a file and chisel and could be made into gears and other complex parts; however, hand working lacked precision and was a slow and expensive process.

James Watt was unable to have an accurately bored cylinder for his first steam engine, trying for several years until John Wilkinson invented a suitable boring machine in 1774, boring Boulton & Watt's first commercial engine in 1776.

The advance in the accuracy of machine tools can be traced to Henry Maudslay and refined by Joseph Whitworth. That Maudslay had established the manufacture and use of master plane gages in his shop (Maudslay & Field) located on Westminster Road south of the Thames River in London about 1809, was attested to by James Nasmyth who was employed by Maudslay in 1829 and Nasmyth documented their use in his autobiography.

The process by which the master plane gages were produced dates back to antiquity but was refined to an unprecedented degree in the Maudslay shop. The process begins with three plates each given an identification (ex., 1,2 and 3). The first step is to rub plates 1 and 2 together with a marking medium (called bluing today) revealing the high spots which would be removed by hand scraping with a steel scraper, until no irregularities were visible. This would not produce absolutely true plane surfaces but a "ball and socket" fit, as this mechanical fit, like two perfect planes, can slide over each other and reveal no high spots. Next, plate number 3 would be compared and scraped to conform to plate number 1. In this manner plates number 2 and 3 would be identical. Next plates number 2 and 3 would be checked against each other to determine what condition existed, either both plates were "balls" or "sockets". These would then be scraped until no high spots existed and then compared to plate number 1. After repeating this process, comparing and scraping the three plates together, they would automatically generate exact true plane surfaces accurate to within millionths of an inch.

The traditional method of producing the surface gages used an abrasive powder rubbed between the plates to remove the high spots, but it was Whitworth who contributed the refinement of replacing the grinding with hand scraping. Sometime after 1825 Whitworth went to work for Maudslay and it was there that Whitworth perfected the hand scraping of master surface plane gages. In his paper presented to the British Association for the Advancement of Science at Glasgow in 1840, Whitworth pointed out the inherent inaccuracy of grinding due to no control and thus unequal distribution of the abrasive material between the plates which would produce uneven removal of material from the plates.

With the creation of master plane gages of such high accuracy, all critical components of machine tools (i.e., guiding surfaces such as machine ways) could then be compared against them and scraped to the desired accuracy.

The first machine tools offered for sale (i.e., commercially available) were constructed by Matthew Murray in England around 1800. Others, such as Henry Maudslay, James Nasmyth, and Joseph Whitworth, soon followed the path of expanding their entrepreneurship from manufactured end products and millwright work into the realm of building machine tools for sale.

Important early machine tools included the slide rest lathe, screw-cutting lathe, turret lathe, milling machine, pattern tracing lathe, shaper, and metal planer, which were all in use before 1840. With these machine tools the decades-old objective of producing interchangeable parts was finally realized. An important early example of something now taken for granted was the standardization of screw fasteners such as nuts and bolts. Before about the beginning of the 19th century, these were used in pairs, and even screws of the same machine were generally not interchangeable. Methods were developed to cut screw thread to a greater precision than that of the feed screw in the lathe being used. This led to the bar length standards of the 19th and early 20th centuries.

American production of machine tools was a critical factor in the Allies' victory in World War II. Production of machine tools tripled in the United States in the war. No war was more industrialized than World War II, and it has been written that the war was won as much by machine shops as by machine guns.

The production of machine tools is concentrated in about 10 countries worldwide: China, Japan, Germany, Italy, South Korea, Taiwan, Switzerland, USA, Austria, Spain and a few others. Machine tool innovation continues in several public and private research centers worldwide.

Drive Power Sources

"all the turning of the iron for the cotton machinery built by Mr. Slater was done with hand chisels or tools in lathes turned by cranks with hand power". David Wilkinson

Machine tools can be powered from a variety of sources. Human and animal power (via cranks, treadles, treadmills, or treadwheels) were used in the past, as was water power (via water wheel); however, following the development of high-pressure steam engines in the mid 19th century, factories increasingly used steam power. Factories also used hydraulic and pneumatic power. Many small workshops continued to use water, human and animal power until electrification after 1900.

Today most machine tools are powered by electricity; however, hydraulic and pneumatic power are sometimes used, but this is uncommon.

Automatic Control

Machine tools can be operated manually, or under automatic control. Early machines used flywheels to stabilize their motion and had complex systems of gears and levers to control the machine and the piece being worked on. Soon after World War II, the numerical control (NC) machine was developed. NC machines used a series of numbers punched on paper tape or punched cards to control their motion. In the 1960s, computers were added to give even more flexibility to the process. Such machines became known as computerized numerical control (CNC) machines. NC and CNC machines could precisely repeat sequences over and over, and could produce much more complex pieces than even the most skilled tool operators.

Before long, the machines could automatically change the specific cutting and shaping tools that were being used. For example, a drill machine might contain a magazine with a variety of drill bits for producing holes of various sizes. Previously, either machine operators would usually have to manually change the bit or move the work piece to another station to perform these different operations. The next logical step was to combine several different machine tools together, all under computer control. These are known as machining centers, and have dramatically changed the way parts are made.

From the simplest to the most complex, most machine tools are capable of at least partial self-replication, and produce machine parts as their primary function.

Examples

Examples of machine tools are:

- Broaching machine
- Drill press

- Gear shaper
- Hobbing machine
- Hone
- Lathe
- Screw machines
- Milling machine
- Shear (sheet metal)
- Shaper
- Saws
- Planer
- Stewart platform mills
- Grinding machines
- Multitasking machines (MTMs)—CNC machine tools with many axes that combine turning, milling, grinding, and material handling into one highly automated machine tool

When fabricating or shaping parts, several techniques are used to remove unwanted metal. Among these are:

- Electrical discharge machining
- Grinding (abrasive cutting)
- Multiple edge cutting tools
- Single edge cutting tools

Other techniques are used to *add* desired material. Devices that fabricate components by selective *addition* of material are called rapid prototyping machines.

Machine Tool Manufacturing Industry

The worldwide market for machine tools was approximately $81 billion in production in 2014 according to a survey by market research firm Gardner Research. The largest producer of machine tools was China with $23.8 billion of production followed by Germany and Japan at neck and neck with $12.9 billion and $12.8 billion respectively. South Korea and Italy rounded out the top 5 producers with revenue of $5.6 billion and $5 billion respectively.

Shaper

A shaper is a type of machine tool that uses linear relative motion between the workpiece and a single-point cutting tool to machine a linear toolpath. Its cut is analogous to that of a lathe, except that it is (archetypally) linear instead of helical.

Shaper tool slide, clapper box and cutting tool

A wood shaper is a similar woodworking tool, typically with a powered cutting head and manually fed workpiece, usually known simply as a *shaper* in North America and *spindle moulder* in the UK.

Shaper with boring bar setup to allow cutting of internal features, such as keyways, or even shapes that might otherwise be cut with wire EDM.

A metalworking shaper is somewhat analogous to a metalworking planer, with the cutter riding a ram that moves relative to a stationary workpiece, rather than the workpiece moving beneath the cutter. The ram is typically actuated by a mechanical crank inside the column, though hydraulically actuated shapers are increasingly used. Adding axes of motion to a shaper can yield helical toolpaths, as also done in helical planing.

Types

Shapers are mainly classified as standard, draw-cut, horizontal, universal, vertical, geared, crank, hydraulic, contour and traveling head, with a horizontal arrangement most common. Vertical shapers are generally fitted with a rotary table to enable curved surfaces to be machined (same idea as in helical planing). The vertical shaper is essentially the same thing as a slotter (slotting machine), although technically a distinction can be made if one defines a true vertical shaper as a machine whose slide can be moved from the vertical. A slotter is fixed in the vertical plane.

Small shapers have been successfully made to operate by hand power. As size increases, the mass of the machine and its power requirements increase, and it becomes necessary to use a motor or other supply of mechanical power. This motor drives a mechanical arrangement (using a pinion gear, bull gear, and crank, or a chain over sprockets) or a hydraulic motor that supplies the necessary movement via hydraulic cylinders.

Operation

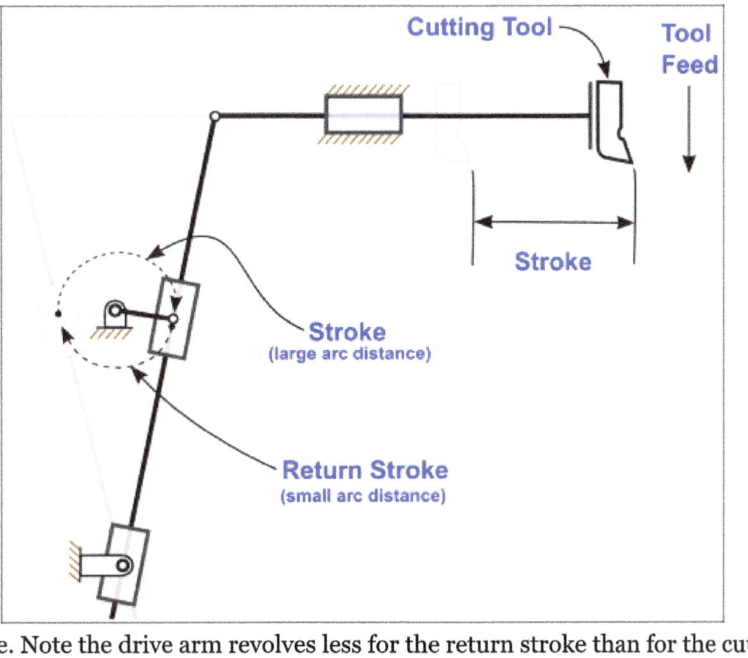

Shaper linkage. Note the drive arm revolves less for the return stroke than for the cutting stroke, resulting in a quicker return stroke and more powerful cutting stroke.

The workpiece mounts on a rigid, box-shaped table in front of the machine. The height of the table can be adjusted to suit this workpiece, and the table can traverse sideways underneath the reciprocating tool, which is mounted on the ram. Table motion may be controlled manually, but is usually advanced by an automatic feed mechanism acting on the feedscrew. The ram slides back and forth above the work. At the front end of the ram is a vertical tool slide that may be adjusted to either side of the vertical plane along the stroke axis. This tool-slide holds the *clapper box* and toolpost, from which the tool can be positioned to cut a straight, flat surface on the top of the workpiece. The tool-slide permits feeding the tool downwards to deepen a cut. This adjustability, coupled with the use of specialized cutters and toolholders, enable the operator to cut internal and external gear tooth

The ram is adjustable for stroke and, due to the geometry of the linkage, it moves faster on the return (non-cutting) stroke than on the forward, cutting stroke. This action is via a slotted link (or *Whitworth* link).

Uses

The most common use is to machine straight, flat surfaces, but with ingenuity and some accessories a wide range of work can be done. Other examples of its use are:

- Keyways in the boss of a pulley or gear can be machined without resorting to a dedicated broaching setup.
- Dovetail slides
- Internal splines and gear teeth.
- Keyway, spline, and gear tooth cutting in blind holes
- Cam drums with toolpaths of the type that in CNC milling terms would require 4- or 5-axis contouring or turn-mill cylindrical interpolation
- It is even possible to obviate wire EDM work in some cases. Starting from a drilled or cored hole, a shaper with a boring-bar type tool can cut internal features that don't lend themselves to milling or boring (such as irregularly shaped holes with tight corners).

History

Roe (1916) credits James Nasmyth with the invention of the shaper in 1836. Shapers were very common in industrial production from the mid-19th century through the mid-20th. In current industrial practice, shapers have been largely superseded by other machine tools (especially of the CNC type), including milling machines, grinding machines, and broaching machines. But the basic function of a shaper is still sound; tooling for them is minimal and very cheap to reproduce; and they are simple and robust in construction, making their repair and upkeep easily achievable. Thus they are still popular in many machine shops, from jobbing shops or repair shops to tool and die shops, where only one or a few pieces are required to be produced and the alternative methods are cost- or tooling-intensive. They also have considerable retro appeal to many hobbyist machinists, who are happy to obtain a used shaper or, in some cases, even to build a new one from scratch.

Planer (Metalworking)

A planer is a type of metalworking machine tool that uses linear relative motion between the workpiece and a single-point cutting tool to cut the work piece. A planer is similar to a shaper, but larger, and with workpiece moving, whereas in a shaper the cutting tool moves.

A typical planer

Applications

Linear Planing

The most common applications of planers and shapers are linear-toolpath ones, such as:

- Generating accurate flat surfaces. (While not as precise as grinding, a planer can remove a tremendous amount of material in one pass with high accuracy.)
- Cutting slots (such as keyways).
- It is even possible to do work that might now be done by wire EDM in some cases. Starting from a drilled or cored hole, a planer with a boring-bar type tool can cut internal features that don't lend themselves to milling or boring (such as irregularly shaped holes with tight corners).

Helical Planing

Although the archetypal toolpath of a planer is linear, helical cutting can be accomplished by coupling the table's linear motion to simultaneous rotation. The helical planing idea is similar to both helical milling and single-point screw cutting.

Current Usage

Planers and shapers are now obsolescent, because other machine tools (such as milling machines, broaching machines, and grinding machines) have mostly eclipsed them as the tools of choice for doing such work. However, they have not yet disappeared from the metalworking world. Planers are used by smaller tool and die shops within larger production facilities to maintain and repair large stamping dies and plastic injection molds. Additional uses include any other task where an abnormally large (usually in the range of 4'×8' or more) block of metal must be squared when a (quite massive) horizontal grinder or floor mill is unavailable, too expensive, or otherwise impractical in a given situation. As usual in the selection of machine tools, an old machine that is in hand, still works, and is long since paid-for has substantial cost advantage over a newer machine that would need to be purchased. This principle easily explains why "old-fashioned" techniques often have a long period of gradual obsolescence in industrial contexts, rather than a sharp drop-off of prevalence such as is seen in mass-consumer technology fashions.

Configurations and Sizes

There are two types of planers for metal: double-housing and open-side. The double-housing variety has vertical supports on both sides of its long bed; the open-side variety has a vertical support on only one side, allowing the workpiece to extend beyond the bed. Metal planers can vary in size from a table size of 30"×72" to 20'×62', and in weight from around 20,000 lbs to over 1,000,000 lbs.

History

Early planing ideas are known to have been underway in France in the 1750s. In the late 1810s, a variety of pioneers in various British shops (including James Fox, George Rennie, Matthew Murray, Joseph Clement, and Richard Roberts) developed the planer into what we today would call a machine tool. The exact details have been contentious and will probably never be known, because the development work being done in various shops was undocumented for various reasons (partially because of proprietary secrecy, and also simply because no one was taking down records for posterity). Roe (1916) provides a short chapter that tells the story as thoroughly as he was able to discover it.

Grinding Machine

A grinding machine, often shortened to grinder, is any of various power tools or machine tools used for grinding, which is a type of machining using an abrasive wheel as the cutting tool. Each grain of abrasive on the wheel's surface cuts a small chip from the workpiece via shear deformation.

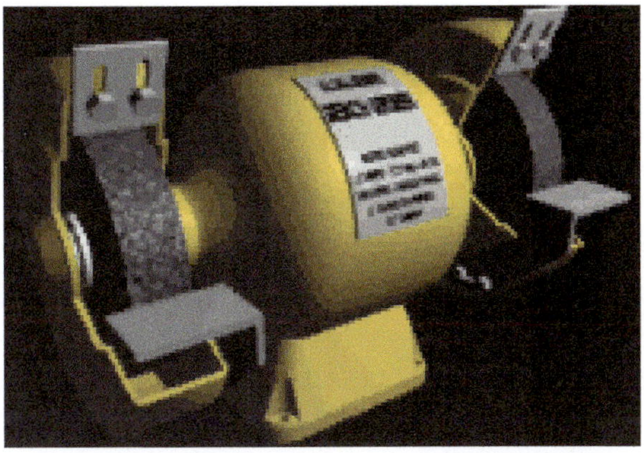

Rotating abrasive wheel on a bench grinder.

Grinding is used to finish workpieces that must show high surface quality (e.g., low surface roughness) and high accuracy of shape and dimension. As the accuracy in dimensions in grinding is on the order of 0.000025 mm, in most applications it tends to be a finishing operation and removes comparatively little metal, about 0.25 to 0.50 mm depth. However, there are some roughing applications in which grinding removes high volumes of metal quite rapidly. Thus, grinding is a diverse field.

Pedal-powered grinding machine, Russia, 1902.

Introduction

The grinding machine consists of a bed with a fixture to guide and hold the work piece, and a power-driven grinding wheel spinning at the required speed. The speed is determined by the wheel's diameter and manufacturer's rating. The grinding head can travel across a fixed work piece, or the work piece can be moved while the grind head stays in a fixed position.

Fine control of the grinding head or table position is possible using a vernier calibrated hand wheel, or using the features of numerical controls.

Grinding machines remove material from the work piece by abrasion, which can generate substantial amounts of heat. To cool the work piece so that it does not overheat and go outside its tolerance, grinding machines incorporate a coolant. The coolant also benefits the machinist as the heat generated may cause burns. In high-precision grinding machines (most cylindrical and surface grinders), the final grinding stages are usually set up so that they remove about 200 nm (less than 1/10000 in) per pass - this generates so little heat that even with no coolant, the temperature rise is negligible.

Types

A surface grinder.

These machines include the:

- Belt grinder, which is usually used as a machining method to process metals and other materials, with the aid of coated abrasives. Sanding is the machining of wood; grinding is the common name for machining metals. Belt grinding is a versatile process suitable for all kind of applications like finishing, deburring, and stock removal.
- Bench grinder, which usually has two wheels of different grain sizes for roughing and finishing operations and is secured to a workbench or floor stand. Its uses include shaping tool bits or various tools that need to be made or repaired. Bench grinders are manually operated.
- Cylindrical grinder, which includes both the types that use centers and the centerless types. A cylindrical grinder may have multiple grinding wheels. The workpiece is rotated and fed past the wheel(s) to form a cylinder. It is used to make precision rods, tubes, bearing races, bushings, and many other parts.

A cylindrical grinder.

- Surface grinder which includes the wash grinder. A surface grinder has a "head" which is lowered to a workpiece which is moved back and forth under the grinding wheel on a table that typically has a controllable permanent magnet for use with magnetic stock but can have a vacuum chuck or other fixturing means. The most common surface grinders have a grinding wheel rotating on a horizontal axis cutting around the circumference of the grinding wheel. Rotary surface grinders, commonly known as "Blanchard" style grinders, have a grinding head which rotates the grinding wheel on a vertical axis cutting on the end face of the grinding wheel, while a table rotates the workpiece in the opposite direction underneath. This type of machine removes large amounts of material and grinds flat surfaces with noted spiral grind marks. It can also be used to make and sharpen metal stamping die sets, flat shear blades, fixture bases or any flat and parallel surfaces. Surface grinders can be manually operated or have CNC controls.

- Tool and cutter grinder and the D-bit grinder. These usually can perform the minor function of the drill bit grinder, or other specialist toolroom grinding operations.

- Jig grinder, which as the name implies, has a variety of uses when finishing jigs, dies, and fixtures. Its primary function is in the realm of grinding holes and pins. It can also be used for complex surface grinding to finish work started on a mill.

- Gear grinder, which is usually employed as the final machining process when manufacturing a high-precision gear. The primary function of these machines is to remove the remaining few thousandths of an inch of material left by other manufacturing methods (such as gashing or hobbing).

- Die grinder, which is a high-speed hand-held rotary tool with a small diameter grinding bit. They are typically air driven (using compressed air), but can be driven with a small electric motor directly or via a flexible shaft.

References

- Thomson, Ross (2009), Structures of Change in the Mechanical Age: Technological Invention in the United States 1790-1865, Baltimore, MD: The Johns Hopkins University Press, ISBN 978-0-8018-9141-0
- Herman, Arthur. Freedom's Forge: How American Business Produced Victory in World War II, pp. 87, 112, 121, 146-50, 161, Random House, New York, NY. ISBN 978-1-4000-6964-4.
- Parker, Dana T. Building Victory: Aircraft Manufacturing in the Los Angeles Area in World War II, pp. 5, 7-8, Cypress, CA, 2013. ISBN 978-0-9897906-0-4.
- "MAS.863/4.140J-P7". fab.cba.mit.edu. Massachusetts Institute of Technology. Retrieved 2016-08-22.
- Weppelmann, E; Brogni, J (March 2014), "A breakthrough in power skiving", Gear Production: A Supplement to Modern Machine Shop: 7–12, retrieved 2014-03-11.

6

Packaging Engineering: An Overview

Packaging engineering is a wide topic, from design conceptualization to product placement. Some of the aspects of packaging engineering that have been dealt with in this text are active packaging, label dispenser, luxury packaging, mandatory labelling and shelf-ready packaging. This section has been carefully written to provide an easy understanding of the varied facets of packaging engineering.

Packaging Engineering

Testing modified atmosphere in a plastic bag of carrots

Packaging engineering, also package engineering, packaging technology and packaging science, is a broad topic ranging from design conceptualization to product placement. All steps along the manufacturing process, and more, must be taken into account in the design of the package for any given product. Package engineering includes industry-specific aspects of industrial engineering, marketing, materials science, industrial design and logistics. Packaging engineers must interact with research and development, manufacturing, marketing, graphic design, regulatory, purchasing, planning and so on. The package must sell and protect the product, while maintaining an efficient, cost-effective process cycle.

Engineers develop packages from a wide variety of rigid and flexible materials. Some materials have scores or creases to allow controlled folding into package shapes (sometimes resembling origami). Packaging involves extrusion, thermoforming, molding and other processing technologies. Packages are often developed for high speed fabrication, filling, processing, and shipment. Packaging engineers use principles of structural analysis and thermal analysis in their evaluations.

Military shipping container being drop tested

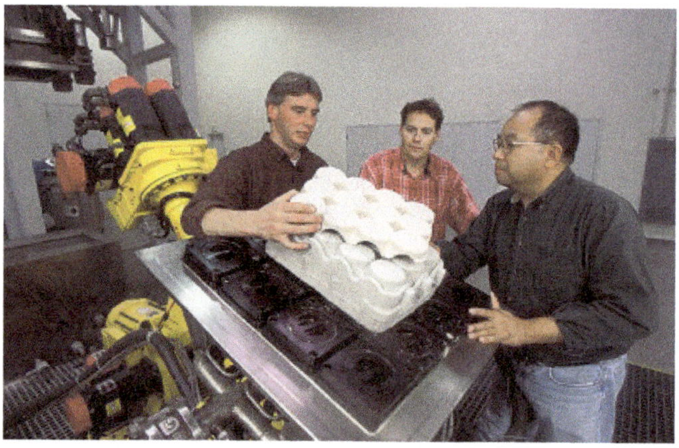

Engineers developing methods of molding packaging components from renewable resources such as straw

Contract Packaging Engineering

Contract Packaging Engineering also known as Staffing Packaging Engineering, is a broad subgroup within the Packaging Engineering Field. Contract Packaging Engineers typically work under the jurisdiction of a written legal arrangement between a company and the engineer or consulting company. Contrary to a full-time engineer, Contract Packaging Engineers work per project and/or predetermined time stated within the contract. Companies typically hire Contract Packaging Engineers for at least one of the three reasons listed below:

1. Company has a need to temporarily staff an open position
2. Company has a desire to hire a candidate on a probationary contract
3. Company is in need of an expert to assist in the execution or management of a project

Packaging Engineering Consultants, similar to Contract Packaging Engineers, are individuals with high level expertise and experience in a packaging engineering area who are hired to manage projects, develop new processes and evolve products.

Education

Some packaging engineers have backgrounds in other science, engineering, or design disciplines while some have college degrees specializing in this field.

Formal packaging programs might be listed as package engineering, packaging science, packaging technology, etc. BE, BS, MS, M.Tech and PhD programs are available. Students in a packaging program typically begin with generalized science, business, and engineering classes before progressing into industry-specific topics such as shelf life stability, corrugated box design, cushioning, engineering design, labeling regulations, project management, food safety, robotics, RFID tags, quality management, package testing, packaging machinery, tamper-evident methods, recycling, computer-aided design, etc.

History

In 1952, Michigan State University became the first university in the USA to offer a degree in Packaging. Faculty and staff are widely published in packaging and other technical subjects.

The University of Wisconsin-Stout is one of only a few schools in the United States (and the only school in the UW System) that offer a B.S. degree in Packaging. The program combines general education with technical and professional studies. The packaging program involves the use of materials, methods, design concepts and machinery to develop and produce the packages that protect and preserve a product, help market the product and instruct the consumer in its proper use. The UW-Stout program offers the opportunity to apply theory to real problems. Through laboratory and co-op work experiences, you will apply the principles of science, mathematics and communications skills.

Throughout its history, UW-Stout has developed technical programs to meet the needs of business and industry. These programs have focused upon the direct application of technical knowledge to the solution of practical problems. The UW-Stout program leads to interesting, challenging, and rewarding careers in package printing; business and sales; foods and packaging; package graphic design; manufacturing and quality management; or package design, research and development.

Rutgers University offers the nation's only packaging program housed in an engineering school, since 1965. Unique from other engineering disciplines in its dynamic, the Packaging Engineering Concentration prepares engineering students by drawing heavily from each of the chemical, industrial, materials and mechanical engineering fields. It requires a strong background in mathematics, chemistry, physics, and computers. The Rutgers Packaging Engineering Program also offers an 18 credit Graduate Packaging Engineering Certificate.

Clemson University offers a bachelor's and masters degrees in Packaging Science, in addition to a PhD in Food Technology. The undergraduate program at Clemson requires student's to take at least one six-month co-op for the degree.

Rochester Institute of Technology's Packaging Science program was first established in 1972. It is an interdisciplinary degree that leads to either a bachelors or masters of science. RIT requires all Packaging Science students to complete a 6-month internship/Co-op.

Central Philippine University in Iloilo City in the Philippines in 2006, is the first to offer (and considered as the first in Asia) in the Philippines a (5-6 year) bachelor's degree in Packaging Engineering. It hosts the Philippine Center for Packaging Engineering and Technology that provides packaging engineering services and consultancy. University alumnus Dr. Lejo C. Brana, considered the founder of the packaging engineering program of the University, is a United States Hall of Famer in Packaging Engineering.

Indian Institute of Technology (IIT)-Roorkee started M.Tech in Packaging Technology in 2015. It is an interdisciplinary degree that leads to bachelors of technology.

In the United Kingdom, the Packaging Society, formerly the Institute of Packaging, offers the industry-standard *Diploma in Packaging Technology*. In India, The Indian institute of Packaging established in 1966, a national apex body which was set up in 1966 by the packaging and allied industries and the Ministry of Commerce, Government of India offers the 2 years Postgraduate Diploma in packaging as well as Graduate Diploma and certificate courses.It has its centers in Mumbai, Kolkata, chennai, Hyderabad and Delhi.Government Polytechnic Nagpur an autonomous institute run by Maharashtra Government offers 3 years Diploma In Packaging Technology.

Packaging and Labeling

Packaging is the technology of enclosing or protecting products for distribution, storage, sale, and use. Packaging also refers to the process of designing, evaluating, and producing packages. Packaging can be described as a coordinated system of preparing goods for transport, warehousing, logistics, sale, and end use. Packaging contains, protects, preserves, transports, informs, and sells. In many countries it is fully integrated into government, business, institutional, industrial, and personal use.

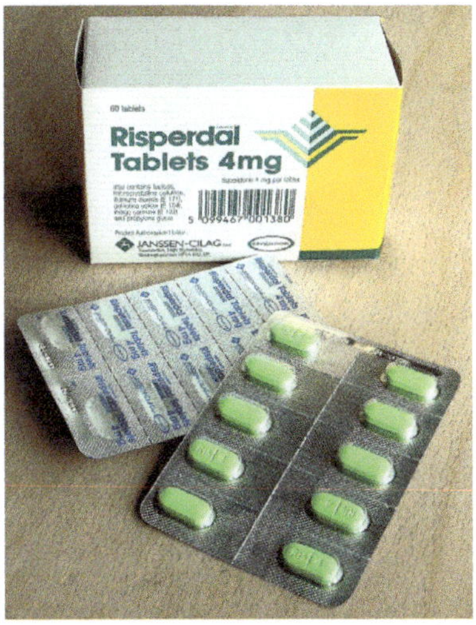

UK Risperdal Tablets 2000 in a blister pack, which was itself packaged in a folding carton made of paperboard

Package labeling (American English) or labelling (British English) is any written, electronic, or graphic communication on the package or on a separate but associated label.

History

Ancient Era

Bronze wine container from the 9th century BCE.

The first packages used the natural materials available at the time: baskets of reeds, wineskins (bota bags), wooden boxes, pottery vases, ceramic amphorae, wooden barrels, woven bags, etc. Processed materials were used to form packages as they were developed: for example, early glass and bronze vessels. The study of old packages is an important aspect of archaeology.

The earliest recorded use of paper for packaging dates back to 1035, when a Persian traveler visiting markets in Cairo noted that vegetables, spices and hardware were wrapped in paper for the customers after they were sold.

Modern Era

Tinning

The use of tinplate for packaging dates back to the 18th century. The manufacture of tinplate was long a monopoly of Bohemia; in 1667 Andrew Yarranton, an English engineer, and Ambrose Crowley brought the method to England where it was improved by ironmasters including Philip Foley. By 1697, John Hanbury had a rolling mill at Pontypool for making "Pontypoole Plates". The method pioneered there of rolling iron plates by means of cylinders enabled more uniform black plates to be produced than was possible with the former practice of hammering.

Tinplate boxes first began to be sold from ports in the Bristol Channel in 1725. The tinplate was shipped from Newport, Monmouthshire. By 1805, 80,000 boxes were made and 50,000 exported. Tobacconists in London began packaging snuff in metal-plated canisters from the 1760s onwards.

Canning

1914 magazine advertisement for cookware with instructions for home canning.

With the discovery of the importance of airtight containers for food preservation by French inventor Nicholas Appert, the tin canning process was patented by British merchant Peter Durand in 1810. After receiving the patent, Durand did not himself follow up with canning food. He sold his patent in 1812 to two other Englishmen, Bryan Donkin and John Hall, who refined the process and product and set up the world's first commercial canning factory on Southwark Park Road, London. By 1813, they were producing the first canned goods for the Royal Navy.

The progressive improvement in canning stimulated the 1855 invention of the can opener. Robert Yeates, a cutlery and surgical instrument maker of Trafalgar Place West, Hackney Road, Middlesex, UK, devised a claw-ended can opener with a hand-operated tool that haggled its way around the top of metal cans. In 1858, another lever-type opener of a more complex shape was patented in the United States by Ezra Warner of Waterbury, Connecticut.

Paper-based Packaging

Packing folding cartons of salt.

Set-up boxes were first used in the 16th century and modern folding cartons date back to 1839. The first corrugated box was produced commercially in 1817 in England. Corrugated (also called pleated) paper received a British patent in 1856 and was used as a liner for tall hats. Scottish-born Robert Gair invented the pre-cut paperboard box in 1890—flat pieces manufactured in bulk that folded into boxes. Gair's invention came about as a result of an accident: as a Brooklyn printer and paper-bag maker during the 1870s, he was once printing an order of seed bags, and the metal ruler, normally used to crease bags, shifted in position and cut them. Gair discovered that by cutting and creasing in one operation he could make prefabricated paperboard boxes.

Commercial paper bags were first manufactured in Bristol, England, in 1844, and the American Francis Wolle patented a machine for automated bag-making in 1852.

20th Century

Packaging advancements in the early 20th century included Bakelite closures on bottles, transparent cellophane overwraps and panels on cartons. These innovations increased processing efficiency and improved food safety. As additional materials such as aluminum and several types of plastic were developed, they were incorporated into packages to improve performance and functionality.

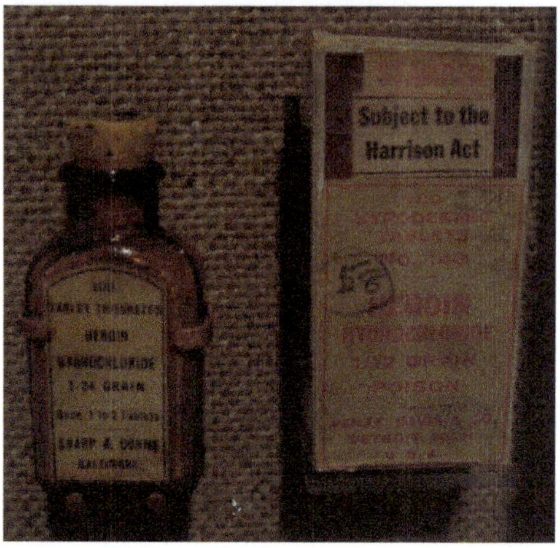

Heroin bottle and carton, early 20th century.

In 1952, Michigan State University became the first university in the world to offer a degree in Packaging Engineering.

In-plant recycling has long been common for producing packaging materials. Post-consumer recycling of aluminum and paper-based products has been economical for many years: since the 1980s, post-consumer recycling has increased due to curbside recycling, consumer awareness, and regulatory pressure.

A pill box made from polyethylene in 1936.

Many prominent innovations in the packaging industry were developed first for military use. Some military supplies are packaged in the same commercial packaging used for general industry. Other military packaging must transport materiel, supplies, foods, etc. under severe distribution and storage conditions. Packaging problems encountered in World War II led to Military Standard or "mil spec" regulations being applied to packaging, which was then designated "military specification packaging". As a prominent concept in the military, mil spec packaging officially came into being around 1941, due to operations in Iceland experiencing critical losses, ultimately attributed to bad packaging. In most cases, mil spec packaging solutions (such as barrier materials, field rations, antistatic bags, and various shipping crates) are similar to commercial grade packaging materials, but subject to more stringent performance and quality requirements.

As of 2003, the packaging sector accounted for about two percent of the gross national product in developed countries. About half of this market was related to food packaging.

The Purposes of Packaging and Package Labels

Packaging and package labeling have several objectives

- Physical protection – The objects enclosed in the package may require protection from, among other things, mechanical shock, vibration, electrostatic discharge, compression, temperature, etc.
- Barrier protection – A barrier to oxygen, water vapor, dust, etc., is often required. Permeation is a critical factor in design. Some packages contain desiccants or oxygen absorbers to help extend shelf life. Modified atmospheres or controlled atmospheres are also maintained in some food packages. Keeping the contents clean, fresh, sterile and safe for the duration of the intended shelf life is a primary function. A barrier is also implemented in cases where segregation of two materials prior to end use is required, as in the case of special paints, glues, medical fluids, etc. At the consumer end, the packaging barrier is broken or measured amounts of material are removed for mixing and subsequent end use.
- Containment or agglomeration – Small objects are typically grouped together in one package for reasons of storage and selling efficiency. For example, a single box of 1000 pencils requires less physical handling than 1000 single pencils. Liquids, powders, and granular materials need containment.
- Information transmission – Packages and labels communicate how to use, transport, recycle, or dispose of the package or product. With pharmaceuticals, food, medical, and chemical products, some types of information are required by government legislation. Some packages and labels also are used for track and trace purposes. Most items include their serial and lot numbers on the packaging, and in the case of food products, medicine, and some chemicals the pack-

aging often contains an expiry/best-before date, usually in a shorthand form. Packages may indicate their construction material with a symbol.

- Marketing – Packaging and labels can be used by marketers to encourage potential buyers to purchase a product. Package graphic design and physical design have been important and constantly evolving phenomena for several decades. Marketing communications and graphic design are applied to the surface of the package and often to the point of sale display. Most packaging is designed to reflect the brand's message and identity.

A single-serving shampoo packet

- Security – Packaging can play an important role in reducing the security risks of shipment. Packages can be made with improved tamper resistance to deter manipulation and they can also have tamper-evident features indicating that tampering has taken place. Packages can be engineered to help reduce the risks of package pilferage or the theft and resale of products: Some package constructions are more resistant to pilferage than other types, and some have pilfer-indicating seals. Counterfeit consumer goods, unauthorized sales (diversion), material substitution and tampering can all be minimized or prevented with such anti-counterfeiting technologies. Packages may include authentication seals and use security printing to help indicate that the package and contents are not counterfeit. Packages also can include anti-theft devices such as dye-packs, RFID tags, or electronic article surveillance tags that can be activated or detected by devices at exit points and require specialized tools to deactivate. Using packaging in this way is a means of retail loss prevention.

- Convenience – Packages can have features that add convenience in distribution, handling, stacking, display, sale, opening, reclosing, using, dispensing, reusing, recycling, and ease of disposal

- Portion control – Single serving or single dosage packaging has a precise amount of contents to control usage. Bulk commodities (such as salt) can be divided into packages that are a more suitable size for individual households. It also aids the control of inventory: selling sealed one-liter bottles of milk, rather than having people bring their own bottles to fill themselves.

Packaging Types

Various types of household packaging for foods

Packaging may be of several different types. For example, a transport package or distribution package can be the shipping container used to ship, store, and handle the product or inner packages. Some identify a consumer package as one which is directed toward a consumer or household.

Packaging may be described in relation to the type of product being packaged: medical device packaging, bulk chemical packaging, over-the-counter drug packaging, retail food packaging, military materiel packaging, pharmaceutical packaging, etc.

It is sometimes convenient to categorize packages by layer or function: "primary", "secondary", etc.

- Primary packaging is the material that first envelops the product and holds it. This usually is the smallest unit of distribution or use and is the package which is in direct contact with the contents.
- Secondary packaging is outside the primary packaging, and may be used to prevent pilferage or to group primary packages together.
- Tertiary packaging is used for bulk handling, warehouse storage and transport shipping. The most common form is a palletized unit load that packs tightly into containers.

These broad categories can be somewhat arbitrary. For example, depending on the use, a shrink wrap can be primary packaging when applied directly to the product, secondary packaging when used to combine smaller packages, or tertiary packaging when used to facilitate some types of distribution, such as to affix a number of cartons on a pallet.

Symbols Used on Packages and Labels

A bar code on a tin of condensed milk

Many types of symbols for package labeling are nationally and internationally standardized. For consumer packaging, symbols exist for product certifications (such as the FCC and TÜV marks), trademarks, proof of purchase, etc. Some requirements and symbols exist to communicate aspects of consumer rights and safety, for example the CE marking or the estimated sign that notes conformance to EU weights and measures accuracy regulations. Examples of environmental and recycling symbols include the recycling symbol, the recycling code (which could be a resin identification code), and the "Green Dot". Food packaging may show food contact material symbols. In the European Union, products of animal origin which are intended to be consumed by humans have to carry standard, oval-shaped EC identification and health marks for food safety and quality insurance reasons.

Bar codes, Universal Product Codes, and RFID labels are common to allow automated information management in logistics and retailing. Country of Origin Labeling is often used. Some products might use QR codes or similar matrix barcodes. Packaging may have visible registration marks and other printing calibration and troubleshooting cues.

Shipping Container Labeling

Technologies related to shipping containers are identification codes, bar codes, and electronic data interchange (EDI). These three core technologies serve to enable the business functions in the process of shipping containers throughout the distribution channel. Each has an essential function: identification codes either relate product information or serve as keys to other data, bar codes allow for the automated input of identification codes and other data, and EDI moves data between trading partners within the distribution channel.

"Print & Apply" corner wrap UCC (GS1-128) label application to a pallet load

Elements of these core technologies include UPC and EAN item identification codes, the SCC-14 (UPC shipping container code), the SSCC-18 (Serial Shipping Container Codes), Interleaved 2-of-5 and UCC/EAN-128 (newly designated GS1-128) bar code symbologies, and ANSI ASC X12 and UN/EDIFACT EDI standards.

Small parcel carriers often have their own formats. For example, United Parcel Service has a MaxiCode 2-D code for parcel tracking.

RFID labels for shipping containers are also increasingly used. A Wal-Mart division, Sam's Club, has also moved in this direction and is putting pressure on its suppliers to comply.

Shipments of hazardous materials or dangerous goods have special information and symbols (labels, placards, etc.) as required by UN, country, and specific carrier requirements. On transport packages, standardized symbols are also used to communicate handling needs. Some are defined in the ASTM D5445 "Standard Practice for Pictorial Markings for Handling of Goods" and ISO 780 "Pictorial marking for handling of goods".

Flammable liquid

Explosives

Fragile material	Keep away from water

Package Development Considerations

Package design and development are often thought of as an integral part of the new product development process. Alternatively, development of a package (or component) can be a separate process, but must be linked closely with the product to be packaged. Package design starts with the identification of all the requirements: structural design, marketing, shelf life, quality assurance, logistics, legal, regulatory, graphic design, end-use, environmental, etc. The design criteria, performance (specified by package testing), completion time targets, resources, and cost constraints need to be established and agreed upon. Package design processes often employ rapid prototyping, computer-aided design, computer-aided manufacturing and document automation.

Palletized and unitized load

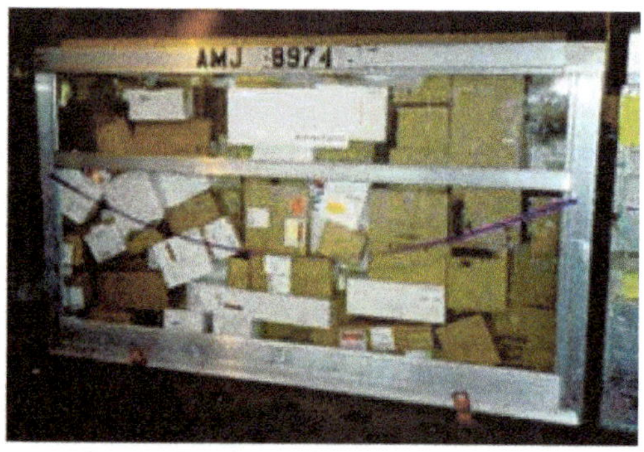

Express air shipment of mixed parcels

Transport packaging needs to be matched to its logistics system. Packages designed for controlled shipments of uniform pallet loads may not be suited to mixed shipments with express carriers.

An example of how package design is affected by other factors is its relationship to logistics. When the distribution system includes individual shipments by a small parcel carrier, the sorting, handling, and mixed stacking make severe demands on the strength and protective ability of the transport package. If the logistics system consists of uniform palletized unit loads, the structural design of the package can be designed to meet those specific needs, such as vertical stacking for a longer time frame. A package designed for one mode of shipment may not be suited to another.

With some types of products, the design process involves detailed regulatory requirements for the packaging. For example, any package components that may contact foods are designated food contact materials. Toxicologists and food scientists need to verify that such packaging materials are allowed by applicable regulations. Packaging engineers need to verify that the completed package will keep the product safe for its intended shelf life with normal usage. Packaging processes, labeling, distribution, and sale need to be validated to assure that they comply with regulations that have the well being of the consumer in mind.

Sometimes the objectives of package development seem contradictory. For example, regulations for an over-the-counter drug might require the package to be tamper-evident and child resistant: These intentionally make the package difficult to open. The intended consumer, however, might be handicapped or elderly and unable to readily open the package. Meeting all goals is a challenge.

Package design may take place within a company or with various degrees of external packaging engineering: independent contractors, consultants, vendor evaluations, independent laboratories, contract packagers, total outsourcing, etc. Some sort of formal project planning and project management methodology is required for all but the simplest package design and development programs. An effective quality management system and Verification and Validation protocols are mandatory for some types of packaging and recommended for all.

Environmental Considerations

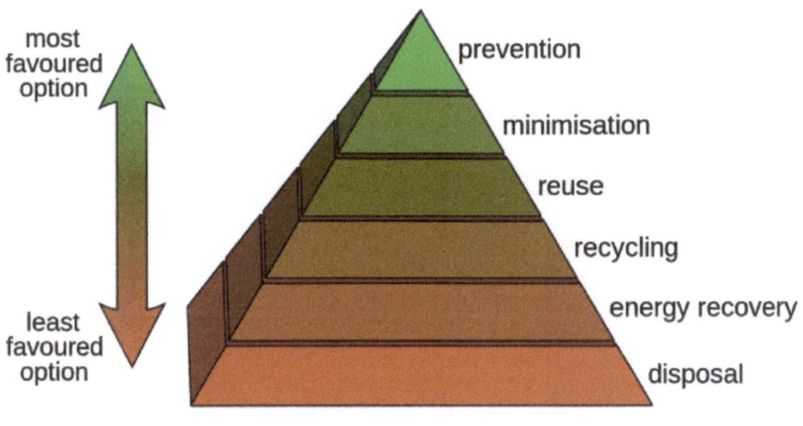

The waste hierarchy

Package development involves considerations of sustainability, environmental responsibility, and applicable environmental and recycling regulations. It may involve a life cycle assessment which considers the material and energy inputs and outputs to the package, the packaged product, the packaging process, the logistics system, waste management, etc. It is necessary to know the relevant regulatory requirements for point of manufacture, sale, and use.

The traditional "three R's" of reduce, reuse, and recycle are part of a waste hierarchy which may be considered in product and package development.

- Prevention – Waste prevention is a primary goal. Packaging should be used only where needed. Proper packaging can also help prevent waste. Packaging plays an important part in preventing loss or damage to the packaged product. Usually, the energy content and material usage of the product being packaged are much greater than that of the package. A vital function of the package is to protect the product for its intended use: if the product is damaged or degraded, its entire energy and material content may be lost.
- Minimization – (also "source reduction") The mass and volume of packaging can be measured and used as criteria for minimizing the package in the design process. Usually "reduced" packaging also helps minimize costs. Packaging engineers continue to work toward reduced packaging.
- Reuse – Reusable packaging is encouraged. Returnable packaging has long been useful (and economically viable) for closed loop logistics systems. Inspection, cleaning, repair and recouperage are often needed. Some manufacturers re-use the packaging of the incoming parts for a product, either as packaging for the outgoing product or as part of the product itself.
- Recycling – Recycling is the reprocessing of materials (pre- and post-consumer) into new products. Emphasis is focused on recycling the largest primary components of a package: steel, aluminum, papers, plastics, etc. Small components can be chosen which are not difficult to separate and do not contaminate recycling operations. Packages can sometimes be designed to separate components to better facilitate recycling.
- Energy recovery – Waste-to-energy and Refuse-derived fuel in approved facilities make use of the heat available from incinerating the packaging components.
- Disposal – Incineration, and placement in a sanitary landfill are undertaken for some materials. Certain US states regulate packages for toxic contents, which have the potential to contaminate emissions and ash from incineration and leachate from landfill. Packages should not be littered.

Development of sustainable packaging is an area of considerable interest to standards organizations, governments, consumers, packagers, and retailers.

Packaging Machines

Choosing packaging machinery includes an assessment of technical capabilities, labor requirements, worker safety, maintainability, serviceability, reliability, ability to integrate into the packaging line, capital cost, floorspace, flexibility (change-over, materials, etc.), energy requirements, quality of outgoing packages, qualifications (for food, pharmaceuticals, etc.), throughput, efficiency, productivity, ergonomics, return on investment, etc.

Beer bottling lines

Packaging machinery can be:

1. purchased as standard, off-the-shelf equipment
2. purchased custom-made or custom-tailored to specific operations
3. manufactured or modified by in-house engineers and maintenance staff

Efforts at packaging line automation increasingly use programmable logic controllers and robotics.

Packaging machines may be of the following general types:

- Accumulating and collating machines
- Blister packs, skin packs and vacuum packaging machines
- Bottle caps equipment, over-capping, lidding, closing, seaming and sealing machines
- Box, case and tray forming, packing, unpacking, closing and sealing machines
- Cartoning machines
- Cleaning, sterilizing, cooling and drying machines
- Coding, printing, marking, stamping, and imprinting machines
- Converting machines
- Conveyor belts, accumulating and related machines
- Feeding, orienting, placing and related machines

- Filling machines: handling dry, powdered, solid, liquid, gas, or viscous products
- Inspecting: visual, sound, metal detecting, etc.
- Label dispenser
- Orienting, unscrambling machines
- Package filling and closing machines
- Palletizing, depalletizing, unit load assembly
- Product identification: labeling, marking, etc.
- Sealing machines: heat sealer
- Slitting machines
- Weighing machines: check weigher, multihead weigher
- Wrapping machines: stretch wrapping, shrink wrap, banding
- Form, fill and seal machines
- Other specialty machinery: slitters, perforating, laser cutters, parts attachment, etc.

Bakery goods shrinkwrapped by shrink film, heat sealer and heat tunnel on roller conveyor	High speed conveyor with stationary bar code scanner for sorting	Label printer applicator applying a label to adjacent panels of a corrugated box.	Robots used to palletize bread
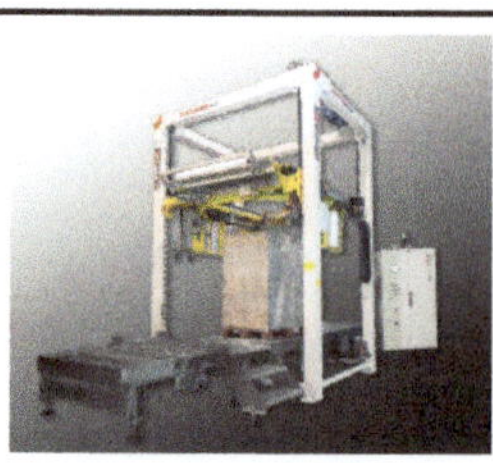			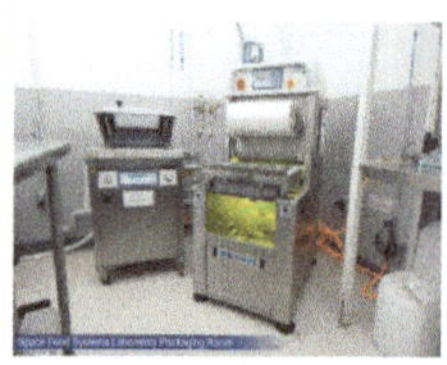
Automatic stretch wrapping machine	Equipment used for making molded pulp components and molding packaging from straw	A semi-automatic rotary arm stretch wrapper	Equipment for thermoforming packages at NASA

Active Packaging

The terms active packaging, intelligent packaging, and smart packaging refer to packaging systems used with foods, pharmaceuticals, and several other types of products. They help extend shelf life, monitor freshness, display information on quality, improve safety, and improve convenience.

The terms are closely related. *Active packaging* usually means having active functions beyond the inert passive containment and protection of the product. *Intelligent* and *smart* packaging usually involve the ability to sense or measure an attribute of the product, the inner atmosphere of the package, or the shipping environment. This information can be communicated to users or can trigger active packaging functions. Programmable matter, smart materials, etc can be employed in packages.

Depending on the working definitions, some traditional types of packaging might be considered as "active" or "intelligent". More often, the terms are used with new technologically advanced systems: microelectronics, computer applications, nanotechnology, etc.

Moisture Control

For many years, desiccants have been used to actively control the water vapor in a closed package. A desiccant is a hygroscopic substance usually in a porous pouch or sachet which is placed inside a sealed package. They have been used to reduce corrosion of machinery and electronics and to extend the shelf life of moisture sensitive foods and drugs.

Corrosion

Corrosion inhibitors can be applied to items to help prevent rust and corrosion. Volatile corrosion inhibitors (VCI) or vapor phase corrosion inhibitors can be provided inside a package in a pouch or can be incorporated in a saturated overwrap of special paper. Many of these are organic salts that condense on the metal to resist corrosion. Some films also have VCI emitting capability.

Films are available with copper ions in the polymer structure, These neutralize the corrosive gas in a package and deter rust.

VCI create a neutral environment in the packaging. It works on the principle of difference in vapour pressure and causes reaction with Metals and Non metals and also with Moisture to render it active to corrosion. There are different forms of VCI's available like Papers, Plastics, HDPE Papers, Oils, Foams, Chips, Aluminum Barrier Foils, Bubble, Emitters etc. that can prevent corrosion at many stages.

Oxygen Control

Oxygen scavengers or oxygen absorbers help remove oxygen from a closed package. Some are small packets or sachets containing powdered iron: as the iron rusts, oxygen is removed from the surrounding atmosphere. Newer systems are on cards or can be built into package films or molded structures. In addition, the physical characteristics of the packaging itself (oxygen transmission rate - OTR) can dictate how effective an oxygen absorber can be, and how long it will stay effective. Packaging with a low OTR will let less oxygen in the closed package through the polymer barrier itself.

Atmosphere

With some products, such as cheese, it has long been common to flush the package with nitrogen prior to sealing: the inert nitrogen is absorbed into the cheese, allowing a tight shrink film package. The nitrogen removes oxygen and "actively" interacts with the cheese to make the package functional.

More recently, other mixtures of gas have been used inside the package to extend the shelf life. The gas mixture depends on the specific product and its degradation mechanisms. Some package components have been developed that incorporate active chemistry to help maintain certain atmospheres in packages.

Oxygen scavengers, carbon dioxide generators, ethanol generators, etc. are available to help keep the atmosphere in a package at specified conditions.

Temperature Monitor

Some temperature indicators give a visual signal that a specified temperature has been exceeded. Others, Time temperature indicators, signal when a critical accumulation of temperature deviation over time has been exceeded. When the mechanism of the indicator is tuned to the mechanism of product degradation, these can provide valuable signals for consumers.

Digital temperature data loggers record the temperatures encountered throughout the shipment. This data can be used to predict product degradation and help determine if the product is suited for normal sale or if expedited sale is required. They also determine the time of the temperature excess: this can be used to direct corrective action.

Thermochromic inks are sometimes used to signal temperature excess or change. Some are reversible while others have a permanent change of color. These can be used alone or with other packaging functions such as barcodes.

The inks can also signal a desired temperature for consumers. For example, one type of beer can has ink that graphically shows when an ideal drinking temperature is achieved.

Controlling Package Temperatures

Self-heating field rations for up to 18 soldiers

For critical vaccines, insulated shipping containers are passive packaging to help control the temperatures fluctuations seen even with a controlled cold chain. In addition, gel packs are often used to actively keep the temperature of the contents within specified acceptable temperature ranges.

Some newer packages have the ability to heat or cool the product for the consumer. These have segregated compartments where exothermic or endothermic reactions provide the desired effect. Self-heating cans are available for several products.

Dispensing

Some packages have closures or other dispensing systems that actively change the contents from a liquid to an aerosol. These are used for products ranging from precision inhalers for medications to spray bottles of household cleaners.

Some dispensing packages for two-part epoxy adhesives do more than passively contain the two components. When dispensed, some packages meter and mix the two components so the adhesive is fully functioning at the point of application.

The ability of a package to fully empty or dispense a viscous liquid is somewhat dependent on the surface energy of the inner walls of the container. The use of superhydrophobic surfaces is useful but can be further improved by using new lubricant-impregnated surfaces.

RFID

Radio-frequency identification chips are becoming more common as smart labels that are used to track and trace packages and unit loads throughout distribution. Newer developments include recording the temperature history of shipments and other intelligent packaging functions.

Security

A variety of security printing methods, security holograms, and specialized labels are available to help confirm that the product in the package is not counterfeit. RFID chips are being used in this application also.

Electronic article surveillance (on the product or on the package) is used to help counter shoplifting.

Microwave Packaging

Metallised films are used as a susceptor for cooking in microwave ovens. These increase the heating capacity and help make foods crisp and brown. Plastic microwavable containers are also used for microwave cooking.

Shock and Vibration

Shock detectors have been available for many years. These are attached to the package or to the product in the package to determine if an excessive shock has been encountered. The mechanisms of these *shock overload* devices have been spring-mass systems, magnets, drops of red dye, and several others.

Recently, digital shock and vibration data loggers have been available to more accurately record the shocks and vibrations of shipment. These are used to monitor critical shipments to determine if extra inspection and calibration is required. They are also used to monitor the types of shocks and vibrations encountered in transit for use in package testing in a laboratory.

Other Developments

Some engineered packaging films contain enzymes, anti-bacterial agents, scavengers, and other active components to help control food degradation and extend shelf life.

Edible films have been developed to allow consumers to eat the package along with the product.

Special packaging has been developed for shipping organs which keeps them alive during extended shipments. The organs are alive and fresh for transplanting.

Regulations

Active packaging is often designed to actively interact with the contents of the package. Thus extra care is often needed for active or smart packagings that are food contact materials.

Food packagers take extra care with some types of active packaging. For example, when the oxygen atmosphere in a package is reduced for extending shelf life, controls for anaerobic bacteria need to be considered. Also when a controlled atmosphere reduces the appearance of food degradation, consumers need to retain a means of determining whether actual degradation is present.

Label Dispenser

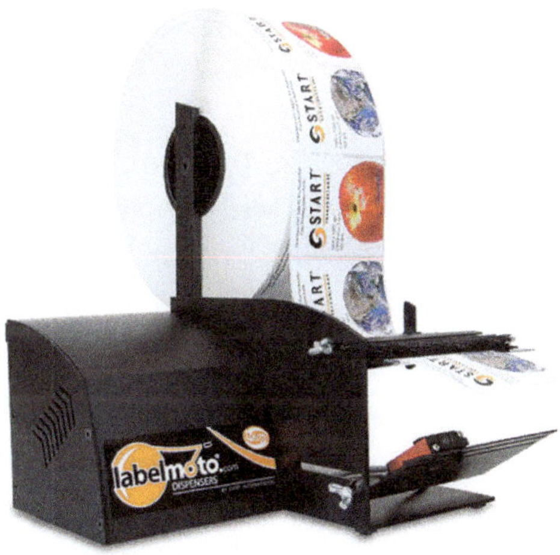

Label dispenser LD6050 from START International

Label dispensers and label applicators are machines built to simplify the process of removing a label from its liner or backing tape. Some are bench-top for dispensing the labels while others include the application of the label to the item (such as a package).

Label dispensers are generally intended to dispense a label to an operator who manually applies the label to the package. They are designed with varying sizes and features which are often specific to the type of label they can dispense and to the degree of automation desired.

Label applicators are usually part of a larger packaging line. They receive the package from a previous automation stage, apply the label, and feed the package to the next stage in the packaging line.

Use

Label dispensers have many uses. Imagine how many items you purchase have labels, almost everything. Those labels were either applied by a machine or by hand, and most likely were peeled from the backing paper using some sort of label dispenser. Some of the more popular are bulk mailing, manufacturing, packaging, food and beverage, fast food, photo labs, and more.

Manual Label Dispenser

Patents on Manual label dispensers go back to the 1920s. They are designed for light-duty use. They are operated by hand and are not automated, but still assist in the process of removing labels from their liners. Many manual label dispensers can dispense multiple rolls at once. Operation is performed by pulling the liner/backing paper around a plate or bar which causes the label to peel away from the backing paper. This happens because the backing paper is usually thinner than the label itself and is also underneath. When the liner is forced around a tight radius the label lifts away and protrudes through the front or top of the dispenser.

Price Guns

A price gun is a tool used in small and large retail outlets to label products with price stickers of varying sizes. While there are a wide variety of pricing gun manufacturers that offer different styles and features, most pricing guns operate in a similar manner i.e. dispensing price labels from a roll.

Using a price gun enables speedy price-marking for products and over-pricing for price reductions for sales or special offer items etc. One of the best benefits of the hand-held price gun is its efficiency and mobility i.e. they are small and lightweight and can be easily taken to the location where the products need to be labelled in store. The gun is held in one hand and by pulling the handle upwards the leading-edge of the roll of labels is pulled through the gun by its internal mechanism and thereby partly ejects one sticker out of the 'mouth' of the gun ready to be applied directly onto the product. The used backing paper is automatically separated from the labels as they are dispensed and that generally comes out the rear of the price gun ready for tearing off as required by the operative.

There are two types of tagging guns, fine fabric and standard. Fine fabric guns are used for delicate materials such as silk and standard is generally used for general garments e.g. coats and knitwear etc. 'Taggers', as they are often referred to, are used widely in the retail industry for applying card material, store branded swing tickets, preprinted with product details and price etc., onto garments of all kinds.

Semi-automatic Label Dispenser

Electric Semi-automatic label dispensers were first patented in the early 1970s. They were originally designed for multiple-row address labels for bulk mailing houses. On average a good mailing house employee could apply approximately 500 labels per hour to envelopes. The label dispenser increased this to over 2,000 per hour. These dispensers advance individual or multiple-row labels and remove them from their lining similar to a manual dispenser, but instead of manually pulling on the liner, label advancement occurs when a trigger on the dispenser detects the absence of a label, such as when the operator removes the label. The sensor then closes the circuit and engages the motor, dispensing the next label until the sensor once again detects the label which opens the circuit. The first electric dispenser was designed with the limit switch on the left of a 16" wide machine. 4-up multiple-row labels were loaded into the machine and once activated would advance one row of labels. The operator would take the labels from right to left, so that when the leftmost label was taken, the next row advanced, automatically providing a constant supply of labels to apply. The labels are also peeled without the natural curl that will happen when pulled from the backing paper with fingers. Also, only one hand was needed to take the label, the other hand could be used to move the material the label was being applied to.

Semi-automatic label dispensers are often built to withstand industrial conditions.

Label Applicators

Label applicators are fully automated and can range from simple slower speed models to large machines capable of applying hundreds or even thousands of labels per minute.

Applicators advance the label stock over the peeler plate until a portion of the label, called the "flag," is extended into the path of the oncoming package. When the package engages the label flag, the label web is advanced to match the speed of the package and label is either tamped or wiped on to assure adhesion. Proper alignment of the label on the package depends on sensors that sense the location/orientation of the package and label sensors that detect the location of the label edge. Package sensors can be a variety of Position sensors, often optical sensors or Ultrasonic sensors. Label sensors are usually Photoelectric sensors because they are inexpensive. But clear labels cannot be detected by photoelectric sensors. Capacitive and ultrasonic technologies are used for clear label detection.

Components

- Motor: The motor automates the label dispenser and controls the speed at which labels are dispensed.
- Feed-Roller: This actually is the first patented method used in the semi-automatic label dispenser to advance the label forward. The label material is fed over the peeling edge, then through a feed-roller assembly which pinches the label material between a metal roller and a rubber roller. The metal roller is held in place with bushings pressed into aluminum blocks which hinge on a rod supported between two sideframes. The blocks use springs on each side to hold tension against the rubber roller (driven by a motor), and this also holds the tracking of the label material to the side of the machine with the sensor. Using this method label stock up to 16" width can be advanced and tracked

perfectly. The disadvantage/advantage to this method is waste is not collected, but pushed out through the rear of the machine where it can be fed into a trash receptacle.

- Take-up hub/spool: This can take the place of the feed-roller. The take-up hub consists of a bar and a liner holder/clip, which are connected to the motor on a single side-frame. In operation, they turn in sync with the motor and wind the label liner in order to collect it in one place. As the take-up hub turns, pressure is placed in the liner and it is stretched across the strip plate, where peeling, or separation, occurs.
- Peeler plate: A component of the machine which acts as a separator for the label and its liner. Each label is pulled across the strip plate until it is recognized by a photo sensor or limit switch. Strip plates can be of varying design and material, but they are often made of plastic, metal coil, or aluminum.
- Photodetector or Limit switch: Used to trigger label advancement. Every semi-automatic label dispenser is equipped with one or the other. These detect the absence or presence of a label to facilitate dispensing.
- Limit switch: This is the original method of label detection and can be used for most applications, it can detect a wide variety of labels.
- Photodetector/Electric eye: Is an alternate method which uses a beam of light broken by the label as it passes over/between the sensors. It can be used for many different label types, but transparent material is an issue.

Luxury Packaging

Packaging and display of cosmetics help create an image of luxury

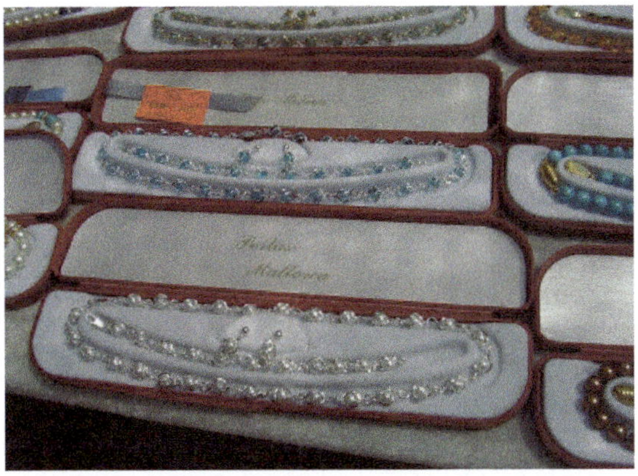

Presentation of jewelry is enhanced by luxury packaging

Luxury and specialty packaging is the design, research, development, and manufacturing of packaging, displays, and for luxury brands. The packaging of a luxury product is part of the brand's image and research shows consumers are willing to spend more on products if the packaging looks appealing and luxurious.

As well as adding to the value of the product, luxury packaging fulfils various other roles; it enhances the image of the brand, increases consumer engagement through personalised packaging, performs a function, creates appeal and diversifies the product.

Unique packaging emphasizes specialized brands

Globally, the luxury packaging market continues to grow, driven by global trends of personalised packaging, attention to sustainability issues, economic and demographic drivers. The luxury packaging market is forecast to grow by 4.4% to 2019, reaching $17.6 billion, and consumption will reach 9.9 billion tons with growth of 3.1%.

Package Development Considerations

Package design and development are often thought of as an integral part of the new product development process. Alternatively, development of a package (or component) can be a separate process, but must be linked closely with the product to be packaged.

Package design starts with the identification of all the requirements: structural design, marketing, shelf life, quality assurance, logistics, legal, regulatory, graphic design, end-use, and environmental. The design criteria, performance (specified by package testing), completion time targets, resources, and cost constraints need to be established and agreed upon. Package design processes often employ rapid prototyping, computer-aided design, computer-aided manufacturing and document automation.

Security

High value products are inviting to theft. Security packaging can be an important consideration to help reduce package pilferage. Authentication technologies can be used to help verify that the expensive luxury product is not among the many counterfeit consumer goods.

Security solutions involve *all phases* of product production, packaging, distribution, logistics, sale, and use. No single solution is considered as "pilfer proof". Often, packaging engineers, logistics engineers, and security professionals have addressed multiple levels of security to reduce the risk of pilfering.

Each situation is unique. Some considerations have included:

- Identifying who a potential thief might be: an internal employee, security guard, truck driver, delivery person, receiver (consignee), organized crime, etc. Engineers usually start with knowing what level of knowledge, materials, tools, etc. might they have.
- Identifying all feasible methods of unauthorized access into a product, package, or system. In addition to the primary means of entry, engineers also consider secondary or "back door" methods.
- Identifying available means of resealing, reclosing, or replacing special seals.
- Using extra strong and secure packaging: A weak or damaged package is an invitation to pilferage.
- Considering unique custom seals and labels (changing regularly because these are subject to counterfeiting)
- Improving the pilfer resistance to make pilfering more difficult, time-consuming, etc.
- Concealing the identity and value of a pilfer able item during distribution. Logistics and packaging professionals do not want bring attention to the item, its package, addresses, names, etc.
- Adding pilfer-evident features to help indicate the existence of pilfering.
- Choosing a logistics provider who can reduce the risks of pilferage.

- Shipping in packages in unit loads with stretch wrap or in intermodal shipping containers with security seals
- Educating people to watch for evidence of pilfering.
- With a corrugated box, using a wider and stronger closure tape, 3-inch or 72 mm, reinforced gummed tape or pressure-sensitive tape.
- Using a tamper-evident tape or seal on packages that leaves a message, warning, or other indication if removed.
- Installing a surveillance system to help identify any suspects.

Design Steps

Some designers consider nine steps to creating the graphic design of luxury packaging, beginning with meeting the client to gain insight into the product. The next step develop first sketches followed by developing the initial layout of the packaging. The fourth step involves refining the idea. The fifth step includes the lettering and typeface. The sixth step covers legalities and laws about how large the obligatory information must be printed, The seventh step involves final adjustments with the eighth step being choosing accents of the packaging. The final step involves designing matching accessories such as boxes or bags to complement the main packaging.

End-use Products

- Cosmetics and fragrances
- Tobacco
- Confectionery
- Premium alcoholic drinks
- Gourmet food and drinks
- Watches and jewellery

Mandatory Labelling

Mandatory labelling or labeling is the requirement of consumer products to state their ingredients or components.

Moral purchasing and problems like allergies are two things which are enabled by labelling. It is mandated in most developed nations, and increasingly in developing nations, especially for food products, *e.g.* "Grade A" meats. With regard to food and drugs, mandatory labelling has been a major battleground between consumer advocates and corporations since the late 19th century.

Because of past scandals involving deceptive labelling, countries like the United States and Canada require most processed foods to have a Nutrition facts label on the label, and the table's formatting and content must conform to strict guidelines. The European Union equivalent is the slightly different Nutrition information table, which may also be supplemented with standardized icons indicating the presence of allergens.

In China, all clothing is labelled with the factory of origin, including telephone and fax numbers, although this information is not available to buyers outside China, who see only a generic Made In China tag.

The development genetically modified food has led to one of the most persistent and divisive debates about the mandatory labelling, which is one of the central topics in genetically modified food controversies. Advocates of such labelling claim that the consumer should make the choice whether to expose themselves to any possible health risk from consuming such foods. Detractors point to studies that conclude genetically modified food is safe, and point out that for many commodity products, the identity of the grower and the custody chain are not known.

Voluntary labelling and co-marketing of products deemed desirable is another matter usually carried out by entirely different means, *e.g.* Slow Food. There has been increased regulatory interest in substantiating these claims, and in some jurisdictions, food labels require regulatory approval before use.

An interesting halfway is those labels that are considered mandatory by one buying population and effectively preclude purchase if they are not there, e.g. kosher, vegan, and the aforementioned GMO-free label now seen on many organic products.

Areas in which mandatory labelling is being discussed include:

- genetically modified food
- spraying of meat with bacteriophages
- in the United States, fresh meat was added to the list of products which require mandatory Country of Origin labels (mCOOL) in 2002, however this was the subject of a challenge in the World Trade Organization by Canada, which Canada won in 2011

Package Testing

Package testing or packaging testing involves the measurement of a characteristic or property involved with packaging. This includes packaging materials, packaging components, primary packages, shipping containers, and unit loads, as well as the associated processes.

Military shipping container being drop tested

Testing measures the effects and interactions of the levels of packaging, the package contents, external forces, and end-use.

It can involve controlled laboratory experiments, subjective evaluations by people, or field testing. Documentation is important: formal test method, test report, photographs, video, etc.

Testing can be a qualitative or quantitative procedure. Package testing is often a physical test. With some types of packaging such as food and pharmaceuticals, chemical tests are conducted to determine suitability of food contact materials. Testing programs range from simple tests with little replication to more thorough experimental designs.

Package testing can extend for the full life cycle. Packages can be tested for their ability to be recycled and their ability to degrade as surface litter, in a sealed landfill or under composting conditions.

Purposes

Packaging testing might have a variety of purposes, such as:

- Determine if, or verify that, the requirements of a specification, regulation, or contract are met
- Decide if a new product development program is on track: Demonstrate proof of concept
- Provide standard data for other scientific, engineering, and quality assurance functions

- Validate suitability for end-use
- Provide a basis for technical communication
- Provide a technical means of comparison of several options
- Provide evidence in legal proceedings: product liability, patents, product claims, etc.
- Help solve problems with current packaging
- Help identify potential cost savings in packaging

Packaging tests can be used for:

- Subjecting packages to stresses and dynamics found in the field
- Reproducing the types of damage to packages and contents found in actual shipments
- Controlling the uniformity of production of packages or components

Importance of Testing

For some types of products, package testing is mandated by regulations: food. pharmaceuticals, medical devices, dangerous goods, etc. This may cover both the design qualification, periodic retesting, and control of the packaging processes. Processes may be controlled by a variety of quality management systems such as HACCP, statistical process control, validation protocols, ISO 9000, etc.

For unregulated products, testing can be required by a contract or governing specification. The degree of package testing can often be a business decision. Risk management may involve factors such as

Testing the ability of packages to resist insect infestation

- costs of packaging
- costs of package testing
- value of contents being shipped
- value of customer's good will
- product liability exposure
- other potential costs of inadequate packaging etc.

With distribution packaging, one vital packaging development consideration is to de-termine if a packaged-product is likely to be damaged in the process of getting to the final customer. A primary purpose of a package is to ensure the safety of a product during transportation and storage. If a product is damaged during this process, then the package has failed to accomplish a primary objective and the customer will either return the product or be unlikely to purchase the product altogether.

Package testing is often a formal part of Project management programs. Packages are usually tested when there is a new packaging design, a revision to a current design, a change in packaging material, and various other reasons. Testing a new packaging design before full scale manufacturing can save time and money.

Laboratory Affiliation

Many suppliers or vendors offer limited material and package testing as a free service to customers. It is common for packagers to partner with reputable suppliers: Many suppliers have certified quality management systems such as ISO 9000 or allow customers to conduct technical and quality audits. Data from testing is commonly shared. There is sometimes a risk that supplier testing may tend to be self-serving and not completely impartial.

Large companies often have their own packaging staff and a package testing and development laboratory. Corporate engineers know their products, manufacturing capabilities, logistics system, and their customers best. Cost reduction of existing products and cost avoidance for new products have been documented.

Another option is to use a paid consultant, Independent contractor, and third-party independent testing laboratory. They are commonly chosen for specialized expertise, for access to certain test equipment, for surge projects, or where independent testing is otherwise required. Many have certifications and accreditations: ISO 9000, ISO/IEC 17025, and various governing agencies.

Procedures

Several standards organizations publish test methods for package testing. Included are:

- International Organization for Standardization, ISO
- ASTM International
- European Committee for Standardization. CEN
- TAPPI
- Military Standards
- ISTA (International Safe Transit Association) etc.

Governments and regulators publish some packaging test methods. There are also many corporate test standards in use. A review of technical literature and patents provides good options to consider for test procedures.

Researchers are not restricted to the use of published standards but can modify existing test methods or develop procedures specific to their particular needs. If a test is conducted with a deviation from a published test method or if a new method is employed, the test report must fully disclose the procedure.

Materials Testing

The basis of packaging design and performance is the component materials. The physical properties, and sometimes chemical properties, of the materials need to be communicated to packaging engineers to aid in the design process. Suppliers publish data sheets and other technical communications that include the typical or average relevant physical properties and the test method these are based upon. Sometimes these are adequate. Other times, additional material and component testing is required by the packager or supplier to better define certain characteristics.

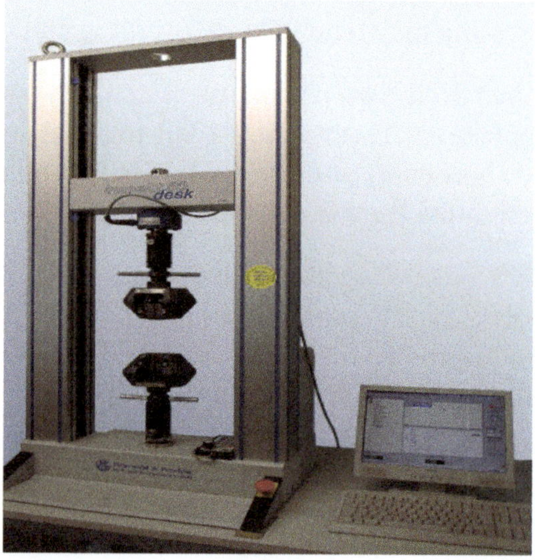

Materials and components are often evaluated on a universal testing machine

When a final package design is complete, the specifications for the component materials needs to be communicated to suppliers. Packaging materials testing is often needed to identify the critical material characteristics and engineering tolerances. These are used to prepare and enforce specifications.

For example, shrink film data might include: tensile strength (MD and CD), elongation, Elastic modulus, surface energy, thickness, Moisture vapor transmission rate, Oxygen transmission rate, heat seal strength, heat sealing conditions, heat shrinking conditions, etc. Average and process capability are often provided. The chemical properties related for use as Food contact materials may be necessary.

Testing with People

Some types of package testing do not use scientific instruments but use people for the evaluation.

The regulations for child-resistant packaging require a test protocol that involves children. Samples of the test packages are given to a prescribed population of children. With specified 50-child panels, a high percentage must be unable to open a test package within 5 minutes. Adults are also tested for their ability to open a child-resistant package.

Consumer packages are often evaluated by focus groups. People evaluate the package features in a room monitored by video cameras. The consumer responses are treated qualitatively for feedback into the new packaging process.

Some food packagers use organoleptic evaluations. People use their senses (taste, smell, etc.) to determine if a package component has tainted the food in the package.

A new package may be evaluated in a test market that uses people to try the packages at home. Consumers have the opportunity to buy a product, perhaps with a coupon or discount. Return postcards or Internet sites provide feedback to package developers. Perhaps the most critical feedback is repeated sales items in the new package. Packaging evaluations are an important part of marketing research.

Legibility of text on packaging and labels is always subjective due to the inherent variations of people. Efforts have been made to help better quantify this by people in a laboratory: still using people for the evaluation but also employing a test apparatus to help reduce variability.

Some laboratory tests are conducted but still result in an observation by people. Some test procedures call for a judgment by test engineers whether or not pre-established acceptance criteria have been met.

Relevant Standards

ASTM D7298 Test Method for Measurement of Comparative Legibility by Means of Polarizing Filter Instrumentation.

ASTM E460 Practice for Determining Effect of Packaging on Food and Beverage Products During Storage

ASTM E619 Practice for Evaluating Foreign Odors in Paper Packaging

ASTM E1870 Test Method for Odor and Taste Transfer from Polymeric Packaging Film

ASTM 2609 Test Method for Odor and Flavor Transfer from Rigid Polymeric Packaging

ISO 16820 Sensory Analysis – Methodology – Sequential Analysis

ISO 5495 Sensory Analysis – Methodology – Paired Comparisons

ISO 13302 Sensory Analysis – Methods for assessing modifications to the flavour of foodstuffs due to packaging

Conditioning, Testing Atmosphere

The environmental conditions of testing are critical. The measured performance of many packages is affected by the conditioning and testing atmospheres. For example, paper based products are strongly affected by their moisture content: Relative humidity needs to be controlled. Plastic products are often strongly affected by temperature.

Conditions of 23 °C (73.4 °F) and 50% relative humidity are common but other standard testing conditions are also published in material and package test standards. Engineering tolerances for the conditions are also specified. Often the package is conditioned to the specified environment and tested under those conditions. This can be in a conditioned room or in a chamber enclosing the test. With some testing, the package is conditioned to a specified environment, then is removed to ambient conditions and quickly tested. The test report needs to state the actual conditions used.

Environmental chamber to simulate temperatures and humidities encountered by packages

Engineers have found it important to know the effects of the full range of expected conditions on package performance. This can be through investigating published technical literature, obtaining supplier documentation, or by conducting controlled tests at diverse conditions.

Relevant Standards

ASTM D4332- Standard Practice for Conditioning Containers, Packages, or Packaging Components for Testing

ASTM E171- Standard Specification for Standard Atmospheres for Conditioning and Testing Flexible Barrier Materials

ASTM F2825 Standard Practice for Climate Stressing of Packaging Systems for Single Parcel Delivery

Degradation of Product

Laboratory tests can help determine the shelf life of a package and its contents under a variety of conditions. This is particularly important for foods, pharmaceuticals, some chemicals, and a variety of products. The testing is usually product specific: the mechanisms of degradation are often different. Exposures to expected and elevated temperatures and humidities are commonly used for shelf life testing. The ability of packaging to control product degradation is frequently a subject of laboratory and field evaluations.

Heat sealer used to prepare bag of lettuce for shelf life testing

Relevant Tests

ASTM E2454 Standard Guide for Sensory Evaluation Methods to Determine the Sensory Shelf -life of Consumer Products

DoD 4140.27M Shelf Life Management Manual, 2000

ISO 11987 Ophthalmic Optics, Contact Lenses, Determination of Shelf Life

Barrier Properties

Many products degrade with exposure to the atmosphere: foods, pharmaceuticals, chemicals, etc. The ability of a package to control the permeation and penetration of gasses is vital for many types of products. Tests are often conducted on the packaging materials but also on the completed packages, sometimes after being subjected to flexing, handling, vibration, or temperature.

Degradation of Packages

Packages can degrade with exposure to temperature, humidity, time, sterlization (steam, radiation, gas, etc.), sunlight, and other environmental factors. For some types of packaging, it is common to test for possible corrosion of metals, polymer degradation, and weather testing of polymers. Several types of accelerated aging of packaging and materials can be accomplished in a laboratory.

Exposure to elevated temperatures accelerates some degradation mechanisms. An Arrhenius equation is often used to correlate certain chemical reactions at different temperatures, based on the proper choice of Q_{10} coefficients.

As with any laboratory testing, validating field trials are important.

Relevant Tests

ASTM D3045 Standard Practice for Heat Aging of Plastics without Load

ASTM F1640 Standard Guide for Packaging Materials for Foods to be Irradiated

ASTM F1980– Standard Guide for Accelerated Aging of Sterile Medical Device Packages

ASTM G151 Standard Practice for Exposing Non-metallic Materials in Accelerated Test Devices that are Laboratory Light Sources

Vacuum Testing

Vacuum chambers are used to test the ability of a package to withstand low pressures. This can be to:

1. Determine the ability of packages to withstand low pressures that might be encountered. this could be in an air shipment or high altitude truck shipment.

2. A laboratory vacuum places controlled stress on a sealed package to test the strength of seals, the tendency for leakage, and the ability to retain sterility.

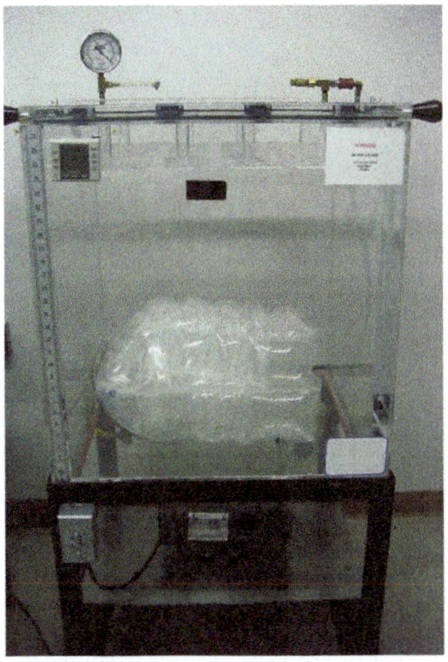

Vacuum chamber for testing leaks in packaging component

Relevant Tests

ASTM D3078- Standard Test Method for Determination of Leaks in Flexible Packaging by Bubble Emission

ASTM D4991- Standard Test Method for Leakage Testing of Empty Rigid Containers by Vacuum Method

ASTM D6653- Standard Test Methods for Determining the Effects of High Altitude on Packaging Systems by Vacuum Method

ASTM D6834- Standard Test Method for Determining Product Leakage from a Package with a Mechanical Pump Dispenser

ASTM E493- Standard Test Methods for Leaks Using the Mass Spectrometer Leak Detector in the Inside-Out Testing Mode

ASTM F2338- Standard Test Method for Nondestructive Detection of Leaks in Packages by Vacuum Decay Method

ASTM F2391- Standard Test Method for Measuring Package and Seal Integrity Using Helium as the Tracer Gas

Shock and Impact

Both primary (consumer) packages and shipping containers have a risk of being dropped or being impacted by other items. Package integrity and product protection are important packaging functions. Tests are conducted to measure the resistance of packages and products to controlled laboratory shock and impact.

Instrumented drop test of cushioned package to measure the transmitted shock

Testing also determines the effectiveness of package cushioning to isolate fragile products from shock. Instrumentation is used to measure the shock transmitted to a cushioned product.

Relevant Tests

ASTM D880- Standard Test Method for Impact Testing for Shipping Containers and Systems

ASTM D1596- Standard Test Method for Dynamic Shock Cushioning Characteristics of Packaging Materials

ASTM D3332- Standard Test Methods for Mechanical-Shock Fragility of Products, Using Shock Machines

ASTM D4003- Standard Test Methods for Programmable Horizontal Impact Test for Shipping Containers and Systems

ASTM D5265- Standard Test Method for Bridge Impact Testing

ASTM D5276- Standard Test Method for Drop Test of Loaded Containers by Free Fall

ASTM D5277- Standard Test Method for Performing Programmed Horizontal Impacts Using an Inclined Impact Tester

ASTM D5487- Standard Test Method for Simulated Drop of Loaded Containers by Shock Machines

ASTM D6344- Standard Test Method for Concentrated Impacts to Transport Packages

ASTM D6537- Standard Practice for Instrumented Package Shock Testing For Determination of Package Performance

Package Insulation

Many packages are used for products that are sensitive to temperature. The ability of insulated shipping containers to protect their contents from exposure to temperature flutuations can be measured in a laboratory. The testing can be of empty containers or of full containers with appropriate jell or ice packs, contents, etc. Ovens, freezers, and environmental chambers are commonly used for this and other types of packaging.

Digital temperature data loggers are used to measure temperatures experienced in different distribution systems. This data is sometimes used to develop unique laboratory test methods for that distribution system.

Relevant Tests

ASTM D3103-Standard Test Method for Thermal Insulation Performance of Distribution Packages

ISTA 7E – Testing Standard for Thermal Transport Packaging Used in Parcel Delivery System Shipment

Thermal Shock

Some packages, particularly glass, can be sensitive to sudden changes in temperature: Thermal shock. One method of testing involves rapid movement from cold to hot water baths, and back.

Relevant Tests

ASTM C149 -Standard Test Method for Thermal Shock Resistance of Glass Containers

MIL-STD-810G METHOD 503.5

Vibration

Vibration is encountered during shipping (vehicle vibration, rough roads, etc.) and movement on conveyors. Potential vibration damage may include:

- fractures and fatigue damage
- loose wires, screw caps, etc.

- bruises on soft products (fruit, etc.)
- surface abrasion etc.

Vibration tester to simulate vibration frequencies at which packaged products are subjected during shipments

The ability of a package to withstand these vibrations and to protect the contents can be measured by several laboratory test procedures. Some allow searching for the particular frequencies of vibration that have potential for damage. Modal testing methodologies are sometimes employed. Others use specified bands of random vibration to better represent complex vibrations measured in field studies of distribution environments.

Relevant Tests

ASTM D999- Standard Test Methods for Vibration Testing of Shipping Containers

ASTM D3580-Standard Test Methods for Vibration (Vertical Linear Motion) Test of Products

ASTM D4728- Standard Test Method for Random Vibration Testing of Shipping Containers

ASTM D5112- Standard Test Method for Vibration (Horizontal Linear Sinusoidal Motion) Test of Products

ASTM D7387- Standard Test Method for Vibration Testing of Intermediate Bulk Containers (IBCs) Used for Shipping Liquid Hazardous Materials (Dangerous Goods)

Compression

Packaging compression tester to determine the stacking strength of packaging

Compression testing relates to stacking or crushing of packages, particularly shipping containers. It usually measures of the force required to crush a package, stack of packages, or a unit load. Packages can be empty or filled as for shipment. A force-deflection curve used to obtain the peak load or other desired points. Other tests use a constant load and measure the time to failure or to a critical deflection.

Dynamic compression is sometimes tested by shock or impact testing with an additional load to crush the test package. Dynamic compression also takes place in stacked vibration testing.

Relevant Tests

ASTM Standard D642 Test Method for Determining Compressive Resistance of Shipping Containers, Components, and Unit Loads.

ASTM Standard D4577 Test Method for Compression Resistance of a Container Under Constant Load

ASTM Standard D7030 Test Method for Short Term Creep Performance of Corrugated Fiberboard Containers Under Constant Load Using a Compression Test Machine

German Standard DIN 55440-1 Packaging Test; compression test; test with a constant conveyance-speed

ISO 12048 Packaging—Complete, filled transport packages—Compression and stacking tests using a compression tester

Large Loads

Large pallet loads, bulk boxes, wooden boxes, and crates can be evaluated by many of the other test procedures previously listed. In addition, some special test methods are available for these larger loads.

Rotational corner drop test of wooden box

Relevant Tests

ASTM D5331- Standard Test Method for Evaluation of Mechanical Handling of Unitized Loads Secured with Stretch Wrap Films

ASTM D5414- Standard Test Method for Evaluation of Horizontal Impact Performance of Load Unitizing Stretch Wrap Films

ASTM D5415- Standard Test Method for Evaluating Load Containment Performance of Stretch Wrap Films by Vibration Testing

ASTM D5416- Standard Test Method for Evaluating Abrasion Resistance of Stretch Wrap Films by Vibration Testing

ASTM D6055- Standard Test Methods for Mechanical Handling of Unitized Loads and Large Shipping Cases and Crates

ASTM D6179- Standard Test Methods for Rough Handling of Unitized Loads and Large Shipping Cases and Crates

ISO 10531- Stability testing of unit loads

Test Protocols for Shipping Containers

Shipping containers are often subjected to sequential tests involving a combination of individual test methods. A variety of standard test schedules or protocols are available for evaluating transport packaging. They are used to help determine the ability of complete and filled shipping containers to various types of logistics systems. Some test the general ruggedness of the shipping container while others have been shown to reproduce the types of damage encountered in distribution. Some base the type and severity of testing on formal studies of the distribution environment: instrumentation, data loggers, and observation. Test cycles with these documented elements better simulate parts of certain logistics shipping environments.

ASTM International

ASTM D4169- Standard Practice for Performance Testing of Shipping Containers and Systems

ASTM D7386- Standard Practice for Performance Testing of Packages for Single Parcel Delivery Systems.

ISO

ISO 4180:2009 Packaging – Complete filled transport packages – General rules for the compilation of performance test schedules

International Safe Transit Association

Procedure 1A: Packaged-Products weighing 150 lb (68 kg) or Less

Procedure 1B: Packaged-Products weighing Over 150 lb (68 kg)

Procedure 1C: Extended Testing for Individual Packaged-Products weighing 150 lb (68 kg) or Less

Procedure 1D: Extended Testing for Individual Packaged-Products weighing Over 150 lb (68 kg)

Procedure 1E: Unitized Loads

Procedure 1G: Packaged-Products weighing 150 lb (68 kg) or Less (Random Vibration)

Procedure 1H: Packaged-Products weighing Over 150 lb (68 kg) (Random Vibration)

Procedure 2A: Packaged-Products weighing 150 lb (68 kg) or Less

Procedure 2B: Packaged-Products weighing over 150 lb (68 kg)

Procedure 2C: Furniture Packages

Procedure 3A: Packaged-Products for Parcel Delivery System Shipments 70 kg (150 lb) or Less (standard, small, flat or elongated)

Procedure 3B: Packaged-Products for Less-Than-Truckload (LTL) Shipment

Procedure 3E: Unitized Loads of Same Product

Procedure 3F: Packaged Products for Distribution Center to Retail Outlet Shipment 100 lb (45 kg)

Procedure 3H: Performance Test for Products or Packaged-Products in Mechanically Handled Bulk Transport Containers

Project 3K: Fast Moving Consumer Goods for the European Retail Supply Chain

Project 4AB: Enhanced Simulation Performance Tests (online test planner)

6-FEDEX-A: FedEx Procedures for Testing Packaged Products Weighing Up to 150 lbs

6-FEDEX-B: FedEx Procedures for Testing Packaged Products Weighing Over 150 lbs

6-SAMSCLUB, Packaged-Products for Sam's Club® Distribution System Shipment

Procedure 7D: Thermal Controlled Transport Packaging for Parcel Delivery System Shipment

ISTA 7E: Testing Standard for Thermal Transport Packaging Used in Parcel Delivery System

Field Trials

Laboratory testing can often help identify shipping container constructions that, in general, should perform well in the field. Of course, laboratory tests cannot fully reproduce the full range of field hazards, their magnitudes, nor their frequency. Field experiments are often conducted to help validate the laboratory testing.

The advantage of laboratory testing is that it subjects replicate packages to identical sets of test sequences: a relatively small number of samples often can suffice. Field hazards, by their nature, are highly variable: thus repeated shipments do not receive the same types or magnitudes of drops, vibrations, kicks, impacts, abrasion, etc. Because of this uncontrolled variability, more replicate sample shipments are often necessary.

Larger scale test markets are used to give additional assurance of performance and acceptability for a new or revised packaged-product. Feedback is carefully obtained and evaluated. Feedback on package performance continues when full production and distribution have been achieved.

Product Requirements

In addition, package testing often relates to the specific product inside the package. Some broad categories of products and special package testing considerations follow:

Food Packaging

Foods categories such as fresh produce, frozen foods, irradiated foods, fresh fish, canned foods, etc. have regulatory requirements and special packaging needs. Package testing often relates to:

Food safety

Compatibility of the package with the food

Migration of material from the packaging to the food

Shelf life

Barrier properties, porosity, package atmosphere, etc

Special quality assurance needs, good manufacturing practices, HACCP, validation protocols, etc

Pharmaceutical Packaging

Packaging for drugs and pharmaceuticals is highly regulated. Special testing needs include:

Safety of drugs and pharmaceuticals

Barrier properties

Shelf life

Compatibility of package with the drugs

Sterility

Tamper resistance, child resistance, etc

Special quality assurance needs, good manufacturing practices, validation protocols, etc

Medical Packaging

Packaging for medical materials, medical devices, health care supplies, etc., have special user requirements and is highly regulated. Barrier properties, durability, visibility, sterility and strength need to be controlled; usually with documented test results for initial designs and for production.

Assurance of sterility and suitability for use are critical. For example, medical devices and products are often sterilized in the package. The sterility must be maintained throughout distribution to allow immediate use by physicians. A series of special packaging tests is used to measure the ability of the package to maintain sterility. Verification and validation protocols are rigidly maintained.

Relevant standards

ASTM F1585 – Guide for Integrity Testing of Porous Medical Packages

ASTM D3078 – Standard Test Method for Detection of Leaks in Flexible Packaging (Bubble)

ASTM F1140 – Standard Test Methods for Internal Pressurization Failure Resistance of Unrestrained Packages

ASTM F1608 – Standard Test Method for Microbial Ranking of Packaging Materials

ASTM F1929 – Standard Test Method for Detecting Seal Strength in Porous Medical Packaging by Dye Penetration

ASTM F2054 – Standard Test Method for Burst Testing of Flexible Package Seals Using Internal Air Pressurization Within Restraining Plates

ASTM F2095 – Standard Test Methods for Pressure Decay Leak Test for Flexible Packages With and Without Restraining Plates

ASTM F2096 – Standard Test Method for Detecting Gross Leaks in Medical Packaging by Internal Pressurization

ASTM F2097 – Standard Guide for Design and Evaluation of Primary Flexible Packaging for Medical Products

ASTM F2228 – Standard Test Method for Non-Destructive Detection of Leaks in Medical Packaging Which Incorporates Pourous Barrier Material by CO2 Tracer Gas

ASTM F2391 – Standard Test Method for Measuring Package and Seal Integrity using Helium as the Tracer Gas

ASTM F3039 - Standard Test Method for Detecting Leaks in Nonporous Packaging or Flexible Barrier Materials by Dye Penetration

EN 868-1 – Packaging materials and systems for medical devices which are to be sterilized. General requirements and test methods (superseded by ISO 11607-1)

ISO 11607-1 – Packaging for terminally sterilized medical devices -- Part 1: Requirements for materials, sterile barrier systems and packaging systems

ISO 11607-2 – Packaging for terminally sterilized medical devices -- Part 2: Validation for Forming, Sealing, and Assembly Processes

Dangerous Goods

Packaging of hazardous materials, or dangerous goods, are highly regulated. There are some material and construction requirements but also performance testing is required. The testing is based on the packing group (hazard level) of the contents, the quantity of material, and the type of container.

Relevant standards

ASTM D4919- Standard Specification for Testing of Hazardous Materials Packaging

ASTM D7387- Standard Test Method for Vibration Testing of Intermediate Bulk Containers (IBCs) Used for Shipping Liquid Hazardous Materials (Dangerous Goods)

ASTM D7760 Standard Guide for Conducting Internal Pressure Tests on United Nations (UN) Packagings

ASTM D7887 Standard Guide for Selection of Substitute, Non-hazardous, Liquid Filling Substances for Packagings Subjected to the United Nations Performance Tests

ASTM D7790: Standard Guide for Preparation of Plastic Packagings Containing Liquids for United Nations (UN) Drop Testing

UN Recommendations on the Transport of Dangerous Goods

ISO 16104 – 2003 Packaging – Transport packaging for dangerous goods – Test methods

Shelf-Ready Packaging

Shelf-ready packaging (SRP) refers to the preparation of a product so that it is delivered to a retailer in a ready-to-sell merchandised unit. Products which come in SRP can be easily placed on the shelf without the need for unpacking or repacking. SRP covers all types of packaging designed for the retail outlet. It is not limited to packaging which goes on the shelf; it also includes sales support mechanisms in all major distribution channels.

SRP is synonymous with RRP Retail-ready packaging, or PAV (prêt-à-vendre). Wal-Mart refers to them as PDQ ("Pretty Darn Quick").

Functional Requirements

The purpose of secondary packaging is to collate and protect the product. In order to call a secondary packaging SRP, additional packaging solutions need to be provided to assure functionality. Experience has shown that on the one hand it is generally easy to define what is not SRP. On the other hand, to define precisely just what SRP is becomes more complicated. Therefore, an agreement on common functional requirements for the design of SRP was implemented by the Efficient Consumer Response (ECR) Europe working group.

The following five SRP requirements are applicable across Europe:

1. easy identification
2. easy open
3. easy dispose
4. easy shelf
5. easy shop

To move products successfully through the supply chain, it is important for a SRP solution to both collate and protect the product, as stated above. A SRP solution which is easy to open but does not protect the product, is not a sufficient SRP.

The different requirements have variable effects throughout the supply chain. The attributes “easy identification”, “easy open”, “easy dispose”, and “easy shelf” have a direct effect on the store staff.

Easy Identification

The requirement “easy identification” supports the store staff in terms of the selection of the right products. All attributes that help in the identification of a product reduce placement error in the stock and in the store and allow a faster filling of the shelves. For this reason, the products on the pallet are arranged and aligned so that different product versions and brands can be seen. In addition, a product must be clearly identified by printing on the SRP. Product, content, versions and expiration date must also be clearly visible on the SRP. The product code should be printed in the form of a barcode, printed to provide easy handling.

Easy Open

All SRP solutions should be easy to open, in order to secure easier and quicker replenishment. Consumers can be negatively affected by overcrowded aisles as well as by out-of-stock situations. To avoid slow and inefficient replenishment, a well designed SRP needs to fulfil the requirement ”easy open”. Another important factor concerning “easy open” is that a SRP solution should be easy to open in the store (without any tools, such as knives), but it should be robust so that it can be transported without damage through the whole supply chain. Additional pictograms with a minimum of text facilitate the handling of SRP and make clear how the SRP is to be handled correctly.

It is important that after opening the SRP, the quality, integrity and appearance of the products are not affected. Furthermore, sharp edges that could pose risks to employees and customers should be avoided. The time required for opening a SRP should contribute to simple and efficient replenishment.

Easy Dispose

In addition, a SRP should be easy to dispose. If the disposal of packaging materials is unwieldy and time-consuming, the benefit generated by SRP are negated by the disadvantages of elaborate disposal. Therefore, SRP should be disposable with little interaction, without assistance, space-saving, and foldable. In order to prevent injury to employees, the tearing of packaging should be avoided. The material used for SRP should be kept as minimal as possible both to save disposal time and to save the environment.

When using re-usable SRP, clear instruction for its return or recycling are necessary.

Easy Shelf

Closely related to the functional requirement "easy open" is the requirement of "easy shelf". Both requirements are intended to facilitate the replenishment of the shelves. To ensure proper replenishment, the SRP needs to be stable and should not lose its original shape. Integrated handles in the SRP can facilitate the placing of goods for the employees in the store. If possible, two SRP should be placed on one shelf to avoid out-of-stock-situations. Furthermore, SRP should be stackable so that the space available on one shelf can be used efficiently.

Easy Shop

So that customer can locate their desired product inside the SRP and easily remove it from the SRP, the requirement "easy shop" needs to be fulfilled. Considering this functional requirement, physical and visual attributes need to be taken into account. The front panel of a SRP should reflect the products included in the SRP. Accordingly, the SRP often includes the brand or manufacturer's logo to ensure rapid recognition and visibility. This makes it easier for the customer to select the desired product, possibly among several variants. SRP should also provide any relevant information from the primary packaging. SRP should improve product identification and thus make a positive contribution to the shopping experience. As a result, SRP should be designed brightly to attract the attention of customers. Individual product variant can be especially enhanced via color printing; plain SRP may reduce the shopping experience. Moreover a customer needs to be able to take the desired product out of the SRP easily and place it back if unwanted. Therefore, products should not be arranged too close in a SRP and should not stick on the tray. After removal of products, the SRP must nevertheless remain stable and still visually appealing for the customer when in a half-empty state.

Depending on the product, type and complexity of the existing SRP, the relative importance of the functional requirements stated above may be different. Often, not all functional requirements can be realized equally by a SRP.

SRP Types

Due to the fact that there are many different products which need to be packed in a SRP, products can be delivered by a variety of solutions.

Shelf Tray

SRP solutions which can be placed in the shelf are predominantly made of corrugated cardboard. In general, these consist of a tray (secondary packaging) and a cover (a lid which protects the product). The cover can be easily separated from the tray by a perforation. Sometimes the cover of a tray is a transparent film which protects the products from mechanical and climatic influences.

Re-usable Plastic Tray

In addition to SRP made of corrugated cardboard, SRP solutions can also be made of re-usable materials such as plastic. Such plastic containers can be re-used directly in the store. Furthermore, if re-use is not possible, the containers can be returned to the producer.

Merchandising Unit – promotional Display, Dolly, Pallet / ½ Pallet / ¼ Pallet

Merchandising units represent another type of SRP solutions. They are used for secondary placement. Normally, promotional goods or fast-moving products are presented in a merchandising unit. An enormous number of products can be placed on a merchandising unit. Therefore, the merchandising unit serves as SRP. Secondary placements can trigger impulse purchases. Often, roll-up palettes, called dollies, are used as a base for merchandising units. Rolling pallets facilitate the handling of merchandising units and increase flexibility.

Importance of Shelf-ready Packaging from the Perspective of Traders

The implementation of SRP differs for discounters and full-line distributors. The implementation rate of full-line distributors averages about 37%, and is therefore considerably lower than the 90% implementation rate for discounters. However, there are also differences among full-line distributors regarding the use of SRP. Basic causes are the respective strategies used in the markets. On the one side, some full-line distributors avoid the use of SRP for image-related and aesthetic reasons. On the other hand, some full-line distributors use SRP extensively because of the ease of handling and lower personnel costs. Apart from the strategy followed for individual markets, the decision whether to use SRP or not depends on the variety of goods provided in the market. The use of SRP in the discount trade has already been mentioned; cost savings can be realized while the best possible presentation of the products is possible. Thus, in some cases revenue could be increased up to 35%, partly because of the use of SRP. In addition to the increase in revenue, SRP simplifies work processes and consequently increases productivity. Activities related to in-store logistics, such as storing, unpacking, labelling products and filling the shelves, account for 70% to 80% of the daily business in the food sector. It is clear that potential savings can be generated through the use of SRP. Thanks to printing on the SRP, selecting stock becomes easier. Due to easier and faster replenishment, the largest potential savings result from this improvement to in-store logistics. Depending on container size, 10 to 60 products can be displayed at once on a shelf. To exploit the full potential savings of SRP, it may be removed from the shelf when the last article has been removed. This leads to gaps in the shelves that can affect the purchasing behaviour of customers. Through the use of an integrated product system, which pushes the remaining products to the front of the SRP whenever one product has been taken from it, shelf gaps can be reduced.

Sustainable Packaging

Sustainable packaging is the development and use of packaging which results in improved sustainability. This involves increased use of life cycle inventory (LCI) and life cycle assessment (LCA) to help guide the use of packaging which reduces the environmental impact and ecological footprint. It includes a look at the whole of the supply chain: from basic function, to marketing, and then through to end of life (LCA) and rebirth. Additionally, an eco-cost to value ratio can be useful The goals are to improve the long term viability and quality of life for humans and the longevity of natural ecosystems. Sustainable packaging must meet the functional and economic needs of the present without compromising the ability of future generations to meet their own needs. Sustainability is not necessarily an end state but is a continuing process of improvement.

Sustainable packaging is a relatively new addition to the environmental considerations for packaging. It requires more analysis and documentation to look at the package design, choice of materials, processing, and life-cycle. This is not just the vague "green movement" that many businesses and companies have been trying to include over the past years. Companies implementing these eco-friendly actions are reducing their carbon footprint, using more recycled materials and reusing more package components. They often encourage suppliers, contract packagers, and distributors to do likewise.

For example, researchers at the Agricultural Research Service are looking into using dairy-based films as an alternative to petroleum-based packaging. Instead of being made of synthetic polymers, these dairy-based films would be composed of proteins such as casein and whey, which are found in milk. The films would be biodegradable and offer better oxygen barriers than synthetic, chemical-based films. More research must be done to improve the water barrier quality of the dairy-based film, but advances in sustainable packaging are actively being pursued.

Environmental marketing claims on packages need to be made (and read) with caution. Ambiguous greenwashing titles such as *green packaging* and *environmentally friendly* can be confusing without specific definition. Some regulators, such as the US Federal Trade Commission, are providing guidance to packagers

Companies have long been reusing and recycling packaging when economically viable. Using minimal packaging has also been a common goal to help reduce costs. Recent years have accelerated these efforts based on social movements, consumer pressure, and regulation. All phases of packaging, distribution, and logistics are included.

Sustainable packaging is no longer focused on just recycling. Just as packaging is not the only eco target, although it is still top of mind for many. Right or wrong, packaging is frequently scrutinized and used as the measure of a company's overall sustainability, even though it may contribute only a small percentage to the total eco impact compared to other things, such as transportation, and water and energy use.

Criteria

The criteria for ranking and comparing packaging based on their sustainability is an active area of development. General guidance, metrics, checklists, and scorecards are being published by several groups.

Government, standards organizations, consumers, retailers, and packagers are considering several types of criteria.

Each organization words the goals and targets a little differently. In general, the broad goals of sustainable packaging are:

1. Functional – product protection, safety, regulatory compliance, etc.
2. Cost effective – if it is too expensive, it is unlikely to be used
3. Support long-term human and ecological health

Specific factors for sustainable design of packaging may include:

- Use of minimal materials – reduced packaging, reduced layers of packaging, lower mass (product to packaging ratio), lower volume, etc.
- Logistics efficiency (through complete life cycle) – cube utilization, tare weight, enablement of efficient transportation, etc.,
- Energy efficiency, total energy content and usage, use of renewable energy, etc.
- Recycled content – as available and functional. For food contact materials, there are special safety considerations, particularly for use of recycled plastics and paper. Regulations are published by each country or region.
- Recyclability – recovery value, use of materials which are frequently and easily recycled, reduction of materials which hinder recyclability of major components, etc.
- Reusable packaging – repeated reuse of package, reuse for other purposes, etc.
- Use of renewable resources in packaging
- Use of biodegradable and compostable materials – when appropriate and do not cause contamination of the recycling stream
- Avoid the use of materials toxic to humans or the environment
- Effects on atmosphere/climate – ozone layer, greenhouse gases (carbon dioxide and methane), volatile organic compounds, etc.
- Water use, reuse, treatment, waste, etc.
- Worker impact: occupational health, safety, clean technology, etc.
- Made from materials healthy throughout the life cycle
- Physically designed to optimize materials and energy
- Use integral design to reduce the use of materials
- Follow the requirements of the 2015 United Nations Climate Change Conference

The chosen criteria are often used best as a basis of comparison for two or more similar packaging designs; not as an absolute success or failure. Such a multi-variable comparison is often presented as a radar chart (spider chart, star chart, etc.).

Benefits

Some aspects of environmentally sound packaging are required by regulators while others are decisions made by each packager. Investors, employees, management, and customers can influence corporate decisions and help set policies. When investors seek to purchase stock, companies known for their positive environmental policy can be attractive. Potential stockholders and investors see this as a solid decision: lower environmental risks lead to more capital at cheaper rates. Companies that highlight their environmental status to consumers, can boost sales as well as product reputation. Going green is often a sound investment that can pay off.

Costs

The process of engineering more environmentally acceptable packages can include consideration of the costs. Some companies claim that their environmental packaging program is cost effective. Some alternative materials that are recycled/recyclable and/or less damaging to the environment can lead to companies incurring increased costs. Though this is common when any product begins to carry the true cost of its production (producer pays, producer responsibility laws, take-back laws). There may be an expensive and lengthy process before the new forms of packaging are deemed safe to the public, and approval may take up to two years. It is important to note here, that for most of the developed world, tightening legislation, and changes in major retailer demand (Walmart's Sustainable Packaging Scorecard for example) the question is no longer "if" products and packaging should become more sustainable, but how-to and how-soon to do it.

Criticism

Efforts toward "greener" packaging are supported in the sustainability community, however, these are often viewed only as incremental steps and not as an end. Some picture a true sustainable steady state economy that may be very different from today's: greatly reduced energy usage, minimal ecological footprint, fewer consumer packaged goods, local purchasing with short food supply chains, little processed foods, etc. Less packaging would be needed in this economy; fewer packaging options would exist; simpler packaging forms may be necessary.

References

- Twede, Diana; Selke, Susan (December 2014). Cartons, Crates & Corrugated Board, handbook of paper & wood packaging technology, Second Edition (2nd ed.). Lancaster PA: DEStech Publications. ISBN 978-1-60595-135-5.
- Soroka (2002) Fundamentals of Packaging Technology, Institute of Packaging Professionals ISBN 1-930268-25-4
- Diana Twede & Susan E.M. Selke (2005). Cartons, crates and corrugated board: handbook of paper and wood packaging technology. DEStech Publications. pp. 41–42, 55–56. ISBN 978-1-932078-42-8.
- Y. Schneider; C. Kluge; U. Weiß; H. Rohm (2010). "Packaging Materials and Equipment". In Barry A. Law, A.Y. Tamime. Technology of Cheesemaking: Second Edition. Wiley-Blackwell. p. 413. ISBN 978-1-4051-8298-0.

- Soroka, W (2008). Illustrated Glossary of Packaging Terms. Institute of Packaging Professionals. p. 3. ISBN 1-930268-27-0.
- Okonkwo, Uché (2007). Luxury Fashion Branding: Trends, Tactics, Techniques. Palgrave Macmillan. ISBN 978-0230590885. Retrieved 2015-01-08.
- Femmely, L. J. (2004). Handbook of Loss Prevention and Crime Prevention (4th ed.). Elsevier Butterworth. p. 281. ISBN 0-7506-7453-9.
- Green, FW (2009), "Export Packaging", in Yam, K L, "Encyclopedia of Packaging Technology", Wiley (published 2010), ISBN 978-0-470-08704-6
- Jedlicka, W, "Packaging Sustainability: Tools, Systems and Strategies for Innovative Package Design", (Wiley, 2008), ISBN 978-0-470-24669-6
- Kunstler, James Howard (2012). Too Much Magic; Wishful Thinking, Technology, and the Fate of the Nation. Atlantic Monthly Press. ISBN 978-0-8021-9438-1.
- McKibben, D, ed. (2010). The Post Carbon Reader: Managing the 21st Centery Sustainability Crisis. Watershed Media. ISBN 978-0-9709500-6-2.
- Speigleman, H, and Sheehan, B. (2010). "Climate Change, Peak Oil, and the End of Waste". In McKibben, D. The Post Carbon Reader: Managing the 21st Centery Sustainability Crisis. Watershed Media. ISBN 978-0-9709500-6-2. CS1 maint: Multiple names: authors list (link)
- Cichello, Simon. "A Guide to Oxygen Absorbers" (PDF). Archived from the original (PDF) on 27 August 2010. Retrieved 12 September 2016.
- Bergman, Robert (2011-12-12). "A New Year: A New Definition Of Luxury". Branding Magazine. Retrieved 2015-01-08.
- Amcor (2014). Sustainability Review 2014. (http://www.amcor.com/sustainability/). Retrieved 10 January 2015.
- Guidance for Industry: Use of Recycled Plastics in Food Packaging: Chemistry Considerations, Contract, US Food and Drug Administrtion, 2006, retrieved 22 Feb 2015
- Title 21, Code of Federal Regulations, Section 176.260 (Pulp from reclaimed fiber) (PDF), US Government, 2009, retrieved 22 Feb 2015

7

Distribution Center: An Integrated Study

Distribution centers are centers that are stocked with products for distribution to wholesalers or consumers directly. Some of the storage mechanisms in the distribution center are automated storage and retrieval system, intermodal container, pallet, pallet racking, bulk boxes and crates. This text will introduce the reader to all the significant aspects of distribution centers.

Distribution Center

Sainsbury's distribution center in Waltham Point, Hertfordshire, United Kingdom.

A distribution center for a set of products is a warehouse or other specialized building, often with refrigeration or air conditioning, which is stocked with products (goods) to be redistributed to retailers, to wholesalers, or directly to consumers. A distribution center is a principal part, the order processing element, of the entire order fulfillment process. Distribution centers are usually thought of as being demand driven. A distribution center can also be called a warehouse, a DC, a fulfillment center, a cross-dock facility, a bulk break center, and a package handling center. The name by which the distribution center is known is commonly based on the purpose of the operation. For example, a "retail distribution center" normally distributes goods to retail stores, an "order fulfillment center" commonly distributes goods directly to consumers, and a cross-dock facility stores little or no product but distributes goods to other destinations.

Distribution centers are the foundation of a supply network, as they allow a single location to stock a vast number of products. Some organizations operate both retail distribution and direct-to-consumer out of a single facility, sharing space, equipment, labor resources, and inventory as applicable.

A typical retail distribution network operates with centers set up throughout a commercial market, with each center serving a number of stores. Large distribution centers for companies such as Wal-Mart serve 50–125 stores. Suppliers ship truckloads of products to the distribution center, which stores the product until needed by the retail location and ships the proper quantity.

Since a large retailer might sell tens of thousands of products from thousands of vendors, it would be impossibly inefficient to ship each product directly from each vendor to each store. Many retailers own and run their own distribution networks, while smaller retailers may outsource this function to dedicated logistics firms that coordinate the distribution of products for a number of companies. A distribution center can be co-located at a logistics center.

Scale

A large distribution center might receive and ship more than ten thousand truckloads each year, with an individual store receiving from only a couple trucks per week up to 20, 30, or more per week. Distribution centers range in size from less than 50,000 square feet (5,000 m^2) to the largest approaching 3 million square feet (3,000,000 m^2).

Storage

Although the primary role of a distribution center is to receive large quantities of products and ship small quantities to individual stores, an important secondary role is storage. Many retailers give priority to having as many items in stock at once as possible. To conserve space, minimize inventory costs, and maximize the variety they offer, the retailer might stock only one or a few items of a particular product. This requires the ability to ship a replacement quickly once an item is sold. By keeping product on hand in the distribution center, the retailer can ship a replacement almost immediately after a product is sold.

Storage Locations and Storage Containers

Goods (products) arrive and are stored in a distribution center in varying types of storage locations and containers suited to the product characteristics and the amount of product to be transported or stored. These types of locations and containers have specific industry-accepted names. Specialized pieces of equipment (material handling equipment, or MHE) are used to handle the various types of containers. The following is a list of some of the names and characteristics of common storage containers:

A fork lift truck moves stacked pallets

- Intermodal containers (shipping containers) are used for the efficient transportation of goods. Standards specify the volume and dimensions of containers to facilitate efficient handling.
- Pallets are one of the most commonly used means to store and move product in a distribution center. There are many specialized devices (material handling equipment or MHE) used to handle pallets - see forklift truck, pallet jack, pallet inverter, and unit load .....Automated Storage and Retrieval Systems (ASRS). Pallets are stored on the floor, may be stacked, and may be stored in pallet racking.
- Gaylords are large single boxes usually connected or attached to a pallet.
- Cases and Cartons are boxes usually containing many items. In distribution centers there is a generally accepted distinction made between the terms "carton" and "case", although both are boxes. Goods are received and stored in cartons, while goods are shipped in cases. A stored carton is called a case once it has been picked or pulled for shipment.
- Totes are reusable containers used to hold and transport goods.

Storage Volume

In addition to shipping quickly, preparing for busy shopping seasons requires retailers to stock up on product ahead of time. For most retailers, the Christmas shopping season is the busiest of the year. Ahead of this time, a distribution center might double the amount of inventory on hand and then draw this level down through the shopping season. This strategy is especially important for imported items. With lead times measured in weeks or months, stocking these products in a distribution center is often the only way to maintain in-stocks at the store. New seasons, holidays, or special promotions may also prompt a retailer to store specific items prior to a large rollout or demand forecast.

Processing

Another way to look at a distribution center is to see it as a production or manufacturing operation. Goods arrive in bulk, they are stored until needed, retrieved, and assembled into shipments. The efficient processing of a distribution center can greatly impact the final price of the product delivered to the end user. Efficient processing not only directly impacts the cost of goods through reduced labor, but it also indirectly impacts the cost of goods through reduced inventory. Inventory represents an investment with its associated investment interest or inventory carrying cost. Reducing the processing time of order processing can directly reduce the amount of inventory necessary to be stocked in the operation.

Costs

The most efficient method of distribution would be to ship a full truckload or railcar directly from the manufacturer to the retailer. The next most efficient method would be to ship a full truckload to a distribution center, unload full pallets of products, and immediately load the pallets onto trucks destined for individual stores. Both of these methods can only be used on very high-volume items. Most products cannot be delivered in this manner, and pallets or even individual boxes must be broken down and divided.

Once a full pallet must be broken apart, the costs of handling the product can increase quickly. Many distribution centers use large sortation systems with miles of conveyor to move products through the facility and into a truck. They may also have automated equipment for de-palletizing and re-palletizing product. Some of the most sophisticated systems can convey product directly into storage racks and then convey out of the racks to trucks, all automatically. With a wide variety of product sizes and weights, these systems are designed to handle a specific range of products. Very large, small, heavy, or light products require varying degrees of manual handling.

As the process of handling involves more steps and becomes increasingly manualized, the cost increases. Storing products instead of receiving and immediately shipping them adds cost. Firms must determine when lost sales due to not having product on the shelves are balanced by increased handling and storage costs.

Distribution Center Organization

All distribution centers have three main areas and may have additional specialized areas. The three main areas are the receiving dock, the storage area, and the shipping dock. In small organizations it is possible for the receiving and shipping functions to occur side by side, but in large centers, separating these areas simplifies the process. Many distribution centers have dedicated dock doors for each store in their shipping area. The receiving area can also be specialized based on the handling characteristics of freight being received, on whether the product is going into storage or directly to a store, or by the type of vehicle delivering the product.

Distribution Center Planning

A number of components go into the overall planning of a distribution center in order to maximize its efficiency. If the distribution center relies on a conveyor system suspended from the ceiling, consideration needs to be given to the weight-bearing capacity of the ceiling joists. If the conveyor system runs along the floor, then consideration needs to be given in the design stage to the placement of columns, particularly as they relate to the flue space between pallet rack frames. Other planning considerations include attention to such areas as slotting, product replenishment, storage media, and power requirements.

Simple Distribution Center Outline

Because many distribution centers service both large and small clients, especially those which store a specific type of service as opposed to those which serve a specific company, roles and departments are generally more complicated. A simple distribution center which serves many clients of a specific theme or type of service may include:

- Goods in: Usually containing specialized container unloading equipment and workers, including pallet wrapping, conveyor belt unloaders (as used on 40 ft shipping containers), forklift drivers, and administrative staff
- Bulk: As a rule, a bulk department controls and ships larger orders or orders that contain only full cartons/boxes. A bulk department includes forklift truck drivers to load containers and wagons, and *man-up* or *combi* forklift trucks to unload full pallets from warehouse racking.
- Break-bulk: Break-bulk (also known as split case) is a lower-capacity version of the *bulk* department. Orders usually contain part boxes or items not requiring pallets. Due to the number of smaller customers a distribution center may serve, a break-bulk department may need more workers than a bulk department. A break-bulk department usually uses trolleys or, for palleted/heavy orders, small electric *PPT* or *walkie low lift* trucks. Items shipped by break-bulk are usually stored in *pick,* which are usually the bottom two *pick-faces* of warehouse racking. A pick-face is the space on such a racking system onto which a pallet can be loaded.
- Export: An export department controls orders which are leaving the country of the distribution center. This department is almost identical in function to a bulk or break-bulk department; however, workers in this department build pallets conforming to different standards and sizes. An export department also uses

different shipping containers or haulage firms.

- Quality assurance: A quality assurance (*QA*) department performs periodic checks of random samples of stock to check quality, including from the warehouse racking, goods in, and returned stock. This department may also take on cycle count duties to find missing stock.
- Administration
- Packing and production: In many distribution centers it is not feasible to store stock in many different packaging styles or quantities, and while it may cost a customer more to do so, many customers, such as supermarkets, prefer their own packaging on stock. Because of this, packing benches are used to take raw items, such as a box of balloons, and pack them at a specific unit quantity, which are then packed into cartons and labeled accordingly for a customer. In many circumstances this may be more inexpensively done at a distribution center than by a customer or client.
- Transportation: Arranges and coordinates shipments in and out of the distribution center.
- Dedicated product departments: Divisions may be based on handling characteristics or storage characteristics, for example, refrigerated and non-refrigerated [meat and produce, frozen, dairy/deli, dry]. Each of these three areas have both shipping and receiving departments as well.

Distribution centers also have a variety of supporting departments, including human resources, maintenance/facilities operations, production control, and accounting.

Distribution Jobs

A distribution center typically has a general manager who manages the facility and typically has a number of department managers who report directly to him/her. Most distribution centers divide staff into two categories, direct labor and indirect labor. Direct labor staff execute the distribution processes, while indirect labor staff support the direct labor staff. Each department is in turn composed of supervisors and warehouse workers. The direct labor jobs of a warehouse can include:

- Unloader - unloads trucks and breaks down pallets as needed, using various pieces of power equipment
- Receiver - inventories and tags unloaded pallets using a mobile cart computer unit and printer
- Hauler - transports received pallets with equipment from the receiving dock to the storage racks

- Putaway driver - puts product into racks with forklift
- Lumper - helps unload shipments
- Replenishment driver - pulls product from the racks and places it into the "pick slot" with forklift
- Order filler - picks product from the "pick slot" by hand and moves with power equipment
- Loader - wraps the order-filled pallets and loads trucks, using equipment

Indirect labor departments and jobs within a warehouse can include:

- Supervision - floor (process) supervision, indirect labor supervision
- Human resources - employment office and employee benefits
- Facilities and housekeeping - maintenance of buildings
- Inventory management - tracking and placement of product
- Quality assurance - inspection and acceptance of incoming and outbound product
- Asset protection - building security and loss prevention
- Safety - insurance of safe operating practices
- Equipment maintenance - electrical, mechanical, and pneumatic maintenance of MHE
- Operations research - Industrial engineering, process improvement, labor standards
- Information technology - support of information systems
- Water Spider - replenishes cardboard, bubble wrap, tape, etc. for warehouse packers

Storage Mechanisms in Distribution Center

Automated Storage and Retrieval System

An automated storage and retrieval system (ASRS or AS/RS) consists of a variety of computer-controlled systems for automatically placing and retrieving loads from defined storage locations. Automated storage and retrieval systems (AS/RS) are typically used in applications where:

- There is a very high volume of loads being moved into and out of storage

- Storage density is important because of space constraints
- No value is added in this process (no processing, only storage and transport)
- Accuracy is critical because of potential expensive damages to the load

Automated 4 aisle miniload warehouse with single mast stacker cranes.

An AS/RS can be used with standard loads as well as nonstandard loads, meaning that each standard load can fit in a uniformly-sized volume; for example the film canisters in the image of the Defense Visual Information Center are each stored as part of the contents of the uniformly sized metal boxes, which are shown in the image. Standard loads simplify the handling of a request of an item. In addition, audits of the accuracy of the inventory of contents can be restricted to the contents of an individual metal box, rather than undergoing a top-to-bottom search of the entire facility, for a single item.

Overview

ASRS with input on ground level

AS/RS systems are designed for automated storage and retrieval of parts and items in manufacturing, distribution, retail, wholesale and institutions. They first originated in the 1960s, initially focusing on heavy pallet loads but with the evolution of the technology the handled loads have become smaller. The systems operate under computerized control, maintaining an inventory of stored items. Retrieval of items is accomplished by specifying the item type and quantity to be retrieved. The computer determines where in the storage area the item can be retrieved from and schedules the retrieval. It directs the proper automated storage and retrieval machine (SRM) to the location where the item is stored and directs the machine to deposit the item at a location where it is to be picked up. A system of conveyors and or automated guided vehicles is sometimes part of the AS/RS system. These take loads into and out of the storage area and move them to the manufacturing floor or loading docks. To store items, the pallet or tray is placed at an input station for the system, the information for inventory is entered into a computer terminal and the AS/RS system moves the load to the storage area, determines a suitable location for the item, and stores the load. As items are stored into or retrieved from the racks, the computer updates its inventory accordingly.

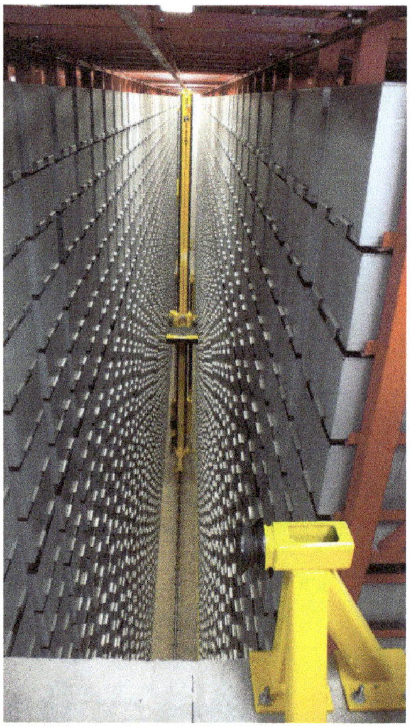

ASRS are also used in libraries allowing for greater safety when consulting and retrieving books.

The benefits of an AS/RS system include reduced labor for transporting items into and out of inventory, reduced inventory levels, more accurate tracking of inventory, and space savings. Items are often stored more densely than in systems where items are stored and retrieved manually.

Within the storage, items can be placed on trays or hang from bars, which are attached to chains/drives in order to move up and down. The equipment required for an AS/RS include a storage & retrieval machine (SRM) that is used for rapid storage and retrieval of material. SRMs are used to move loads vertically or horizontally, and can also move laterally to place objects in the correct storage location.

The trend towards Just In Time production often requires sub-pallet level availability of production inputs, and AS/RS is a much faster way of organizing the storage of smaller items next to production lines.

Material Handling Institute of America (MHIA), the non-profit trade association for the material handling world, and its members have broken AS/RS into two primary segments: Fixed Aisle and Carousels/Vertical Lift Modules (VLMs). Both sets of technologies provide automated storage and retrieval for parts and items, but use different technologies. Each technology has its unique set of benefits and disadvantages. Fixed Aisle systems are characteristically larger systems whereas carousels and Vertical Lift Modules are used individually or grouped, but in small to medium-sized applications.

A fixed-aisle AS/R machine (stacker crane) is one of two main designs: single-masted or double masted. Most are supported on a track and ceiling guided at the top by guide rails or channels to ensure accurate vertical alignment, although some are suspended from the ceiling. The 'shuttles' that make up the system travel between fixed storage shelves to deposit or retrieve a requested load (ranging from a single book in a library system to a several ton pallet of goods in a warehouse system). The entire unit moves horizontally within an aisle, while the shuttles are able to elevate up to the necessary height to reach the load, and can extend and retract to store or retrieve loads that are several positions deep in the shelving. A semi-automated system can be achieved by utilizing only specialized shuttles within an existing rack system.

Automated storage and retrieval system using the highly dynamic TGW Stingray shuttle technology.

Another AS/RS technology is known as shuttle technology. In this technology the horizontal movement is made by independent shuttles each operating on one level of the rack while a lift at a fixed position within the rack is responsible for the vertical movement. By using two separate machines for these two axis the shuttle technology is able to provide higher throughput rates than stacker cranes.

Storage and Retrieval Machines pick up or drop off loads to the rest of the supporting transportation system at specific stations, where inbound and outbound loads are precisely positioned for proper handling.

In addition, there are several types of Automated Storage & Retrieval Systems (AS/RS) devices called Unit-load AS/RS, Mini-load AS/RS, Mid-Load AS/RS, Vertical Lift Modules (VLMs), Horizontal Carousels and Vertical Carousels. These systems are used either as stand-alone units or in integrated workstations called pods or systems. These units are usually integrated with various types of pick to light systems and use either a microprocessor controller for basic usage or inventory management software. These systems are ideal for increasing space utilization up to 90%, productivity levels by 90%, accuracy to 99.9%+ levels and throughput up to 750 lines per hour/per operator or more depending on the configuration of the system.

Advantages

An effective automated storage and retrieval system provides several benefits for supply chain management:

- An efficient AS/RS system helps companies cut expenses by minimizing the amount of unnecessary parts and products in storage, and improving organization of the contents of a warehouse. Due to automated processes, it also allows for more storage space due to high-density storage, narrower aisles, etc.
- Automation reduces labor costs while lowering workforce requirements and increasing safety.
- Modeling and managing the logical representation of the physical storage facilities (e.g. racking, etc.). For example, if certain products are often sold together or are more popular than others, those products can be grouped together or placed near the delivery area to speed up the process of picking, packing and shipping to customers.
- Enabling a seamless link to order processing and logistics management in order to pick, pack, and ship product out of the facility.
- Tracking where products are stocked, which suppliers they come from, and the length of time they are stored. By analyzing such data, companies can control inventory levels and maximize the use of warehouse space. Furthermore, firms are more prepared for the demands and supplies of the market, especially during special circumstances such as a peak season on a particular month. Through the reports generated by an AS/RS system, firms are also able to gather important data that may be put in a model for it to be analyzed.

Vertical Lift Module

VLMs can be built quite high to match the available overhead space in a facility. Multiple units can be places in 'pods' whereby an operator can retrieve items from one unit while the other units are moving. Variants include width, height, load, speed and a control system.

The VLM is a board controlled automated vertical lift module. Inventory within the VLM is stored on front and rear tray locations or rails. When a tray is requested, either by entering a tray number in the built in control pad or by requesting a part through software, an extractor travels vertically between the two columns of trays and pulls the requested tray from its location and brings it to an access point. The operator then picks or replenishes stock and the tray is returned to its home upon confirmation.

VLM systems are sold in numerous configurations, which could be applied in different industries, logistics, as well as office settings. The VLM systems could be customized to fully utilize the height of the facility, even through multiple floors. With the capability of multiple access openings on different floors, the VLM system is able to provide an innovative storage and retrieval solution. The rapid movement of the extractor as well as inventory management software can dramatically increase the efficiency of the picking process. This occurs by simultaneously retrieving and storing trays in multiple units. Unlike large AS/RS systems, which require a complete overhaul of the warehouse or production line, the vertical lift modules are modularized, which can be easily integrated into the existing system, or to be rolled out in gradually over different phases.

Most common applications include: MRO, order picking, consolidation, kitting, parts handling, buffering, inventory storage, WIP, buffer storage, and many more.

VLMs provide floor space savings, increased labor productivity, improved worker ergonomics, and controlled process.

Most VLMs offer dynamic space storage which measures the tray every time it is returned to the unit to optimize space, safety features and some offer tilt tray delivery for increased ergonomic accessibility, and laser pointers which indicate the exact item to be picked on each tray.

Horizontal Carousels

A horizontal carousel is a series of bins which revolve on an oval track. Every bin has shelves which are adjustable to .75" and can be configured for a myriad of standard and special applications. An operator simply inputs a bin number, part number or cell location and the carousel will rotate via the shortest path. Multiple horizontal carousels integrated with pick to light technology and inventory management software (a pod of carousels) are used for order fulfillment.

A wave of orders are sent to the pod. A group of orders are selected to create a batch. The operator simply follows the lights and pick round robin from the carousels and place items in a batch station behind them. Each carousel pre-positions and rotates when picked. By applying the "product to person" principle, operators do not have to move from their position to prepare the order.

When the batch is complete, a new batch is inducted and the process repeated until the wave is complete. Horizontal carousels can save up to 75% of floorspace, increase productivity by 2/3, accuracy levels to 99.9%+ levels and throughput up to 750 lines per hour/operator.

Horizontal carousel systems generally outperform robotic systems for a fraction of the cost. Horizontal carousels are the most cost effective AS/RS system available.

Robotic Inserter/Extractor devices can also be used for horizontal carousels. The robotic device is positioned in the front or rear of up to three horizontal carousels tiered high. The robot grabs the tote required in the order and often replenishes at the same time to speed up throughput. The tote(s) are then delivered to conveyor which routes it to a work station for picking or replenishing. Up to eight transactions per minute per unit can be done. Totes or containers up to 36" x 36" x 36" can be used in a system.

On a simplistic level, horizontal carousels are also often used as "rotating shelving." 'With simple "fetch" command items are brought to the operator and otherwise wasted space is eliminated.

Installed Applications

Installed applications of this technology can be wide ranging. In some libraries, such as at University of Nevada, Reno library, such a system is employed to retrieve books. Still others in use involve retrieval of bicycles from a bicycle tree, as in the case of systems in Japan.

Libraries Using Automated Storage and Retrieval Systems

- University of British Columbia, Vancouver, British Columbia
- Christopher Center, Valparaiso University, Valparaiso, Indiana
- Colgate University, Hamilton, New York
- Cornell University, Ithaca, New York
- North Carolina State University, Raleigh, North Carolina
- University of Chicago, Chicago, Illinois (opening fall 2010)
- University of Louisville, Louisville, Kentucky

- University of Nevada, Las Vegas
- University of Nevada, Reno
- Waseda University, Tokyo (Japan)
- Utah State University, Logan, Utah
- Santa Clara University, Santa Clara, California (robotic retrieval system)
- Sonoma State University, Rohnert Park, California
- California State University Northridge, Northridge, California
- Defense Visual Information Center, March Air Reserve Base, Riverside County, California
- Grand Valley State University, Grand Rapids, Michigan
- Macquarie University, Sydney, Australia (opening 2010)
- Marywood University, Scranton, Pennsylvania
- Chicago State University, Chicago, Illinois
- Georgia Southern University, Statesboro, Georgia

Types

- Vertical Carousels
- Vertical Lift Modules (VLM)
- Horizontal Carousels
- Robotics
- Mini Loads - also known as mini load stackers
- Unit load stackers
- Compact 3D AS/RS

Man-aboard Systems

A man-aboard system can provide significant floorspace savings over manual or fork lift operations but is not truly an AS/RS, as the operation is still manual. Storage system heights are not limited by the reach height of the order picker, as the picker rides along on the platform as it is moved vertically or horizontally to the various storage locations. Shelves or storage cabinets can be stacked as high as floor loading, weight capacity, throughput requirements, and/or ceiling heights will permit. Man-aboard storage and retrieval systems are far and away the most expensive picker-to-stock equipment alternative, but are less expensive than a fully automated system. Aisle-captive storage/retrieval machines reaching heights up to 40 feet cost around $125,000. Hence, there must be enough storage density and/or productivity improvement over cart and tote picking to justify the investment. Also, because vertical travel is slow compared to horizontal travel, typical picking rates in man-aboard operations range between 40 and 250 lines per person-hour. The range is large because there is a wide variety of operating schemes for man-aboard systems. Man-aboard systems are typically appropriate for slow-moving items where space is fairly expensive.

Intermodal Container

A 40-foot (12 m) long shipping container. Each of the eight corners has an essential twistlock fitting for hoisting, stacking, and securing

In 2012 there were over 20 million intermodal containers in the world.

Freight train carrying containers through West Kingman Canyon, Arizona

Making containers *stackable* made loading and transport on large ships feasible and efficient

An intermodal container is a large standardized shipping container, designed and built for intermodal freight transport, meaning these containers can be used across different modes of transport – from ship to rail to truck – without unloading and reloading their cargo. Intermodal containers are primarily used to store and transport materials and products efficiently and securely in the global containerized intermodal freight transport system, but smaller numbers are in regional use as well. These containers are known under a number of names, such as simply container, cargo or freight container, ISO container, shipping, sea or ocean container, container van or (Conex) box, sea or c can.

Intermodal containers exist in many types and a number of standardized sizes, but ninety percent of the global container fleet are so-called *"dry freight"* or *"general purpose"* containers, durable closed steel boxes, mostly of either twenty or forty foot (6 or 12m) standard length. The common heights are 8 feet 6 inches (2.6 m) and 9 feet 6 inches (2.9 m) – the latter are known as High Cube or Hi-Cube containers.

Just like cardboard boxes and pallets, these containers are a means to bundle cargo and goods into larger, unitized loads, that can be easily handled, moved, and stacked, and that will pack tightly in a ship or yard. Intermodal containers share a number of key construction features to withstand the stresses of intermodal shipping, to facilitate their handling and to allow stacking, as well as being identifiable through their individual, unique ISO 6346 reporting mark.

In 2012 there were about 20.5 million intermodal containers in the world of varying types to suit different cargoes. Containers have largely supplanted the traditional break bulk cargo – in 2010 containers accounted for 60% of the world's seaborne trade. The predominant alternative methods of transport carry bulk cargo – either gaseous, liquid or solid – e.g. by bulk carrier or tank ship, tank car or truck. For air freight, the more light-weight IATA-defined unit load device is used.

History

Transferring freight containers on the London, Midland and Scottish Railway (LMS; 1928)

Freight car in railway museum Bochum-Dahlhausen, showing four different UIC-590 pa-containers

In 1975, many containers still featured riveted aluminum sheet and post wall construction, instead of welded, corrugated steel.

By the 1830s, railways on several continents were carrying containers that could be transferred to other modes of transport. The Liverpool and Manchester Railway in the United Kingdom was one of these. "Simple rectangular timber boxes, four to a truck, they were used to convey coal from the Lancashire collieries to Liverpool, where they were transferred to horse-drawn carts by crane." Early versions of standardized containers were used in Europe before World War II. Construction of these containers had a steel frame with wooden walls, floor, roof and doors.

The first international standard for containers was established by the Bureau International des Containers et du Transport Intermodal (B.I.C.) in 1933, and a second one in 1935, primarily for transport between European countries. American containers at this time were not standardized, and these early containers were not yet stackable – neither in the U.S. nor Europe. In November 1932, the first container terminal in the world was opened by the Pennsylvania Rail Road Company in Enola, PA. The development of containerization was created in Europe and the US as a way to revitalize rail companies after the Wall Street Crash of 1929, in New York, and resulting economic collapse and drop in all modes of transport.

In April 1951 at Zürich Tiefenbrunnen railway station the Swiss Museum of Transport and the *Bureau International des Containers* (BIC) held demonstrations of container systems for representatives from a number of European countries, and from the United States. A system was selected for Western Europe, based on the Netherlands' system for consumer goods and waste transportation called *Laadkisten* (lit. "Loading bins"), in use since 1934. This system used roller containers for transport by rail, truck and ship, in various configurations up to 5,500 kg (12,100 lb) capacity, and up to 3.1 by 2.3 by 2 metres (10 ft 2 in × 7 ft 6 $\frac{1}{2}$ in × 6 ft 6 $\frac{3}{4}$ in) in size. This became the first post World War II European railway standard of the International Union of Railways – *UIC-590*, known as "pa-Behälter." It was implemented in the Netherlands, Belgium, Luxembourg, West Germany, Switzerland, Sweden and Denmark.

The use of standardized steel shipping containers began during the late 1940s and early 1950s, when commercial shipping operators and the US military started developing such units. In 1948 the U.S. Army Transportation Corps developed the "Transporter", a rigid, corrugated steel container, able to carry 9,000 pounds (4,100 kg). It was 8 ft 6 in (2.59 m) long, 6 ft 3 in (1.91 m) wide, and 6 ft 10 in (2.08 m) high, with double doors on one end, was mounted on skids, and had lifting rings on the top four corners. After proving successful in Korea, the Transporter was developed into the Container Express (CONEX) box system in late 1952. Based on the Transporter, the size and capacity of the Conex were about the same, but the system was made *modular*, by the addition of a smaller, half-size unit of 6 ft 3 in (1.91 m) long, 4 ft 3 in (1.30 m) wide and 6 ft 10 $\frac{1}{2}$ in (2.10 m) high. CONEXes could be stacked three high, and protected their contents from the elements. By 1965 the US military used some 100,000 Conex boxes, and more than 200,000 in 1967. making this the first worldwide application of intermodal containers.

From 1949 onwards, engineer Keith Tantlinger repeatedly contributed to the development of containers, as well as their handling and transportation equipment. In 1949, while at Brown Trailers Inc. of Spokane, he modified the design of their stressed skin aluminum 30-foot trailer, to fulfil an order of two-hundred 30 by 8 by 8.5 feet (9.14 m

× 2.44 m × 2.59 m) containers that could be stacked two high, for Alaska-based *Ocean Van Lines*. Steel castings on the top corners provided lifting and securing points.

In 1955 trucking magnate Malcom McLean bought Pan-Atlantic Steamship Company, to form a container shipping enterprise, later known as Sea-Land. The first containers were supplied by Brown, where McLean met Keith Tantlinger, and hired him as vice-president of engineering and research. Under the supervision of Tantlinger, a new 35 ft (10.67 m) x 8 ft (2.44 m) x 8 ft 6 in (2.59 m) Sea-Land container was developed, the length determined by the maximum length of trailers then allowed on Pennsylvanian highways. Each container had a frame with eight corner castings that could withstand stacking loads. Tantlinger also designed automatic spreaders for handling the containers, as well as the twistlock mechanism that connects with the corner castings.

Every international shipping container must have a "CSC-Plate"

Two years after McLean's first container ship, the *Ideal X* started container shipping on the U.S. East Coast, Matson Navigation followed suit between California and Hawaii. Just like Pan-Atlantic's containers, Matson's were 8 ft (2.44 m) wide and 8 ft 6 in (2.59 m) high, but due to California's different traffic code, Matson chose to make theirs 24 ft (7.32 m) long. In 1968, McLean began container service to South Vietnam for the US military with great success.

ISO standards for containers were published between 1968 and 1970 by the International Maritime Organization. These standards allow for more consistent loading, transporting, and unloading of goods in ports throughout the world, thus saving time and resources.

The International Convention for Safe Containers is a 1972 regulation by the Inter-governmental Maritime Consultative Organization on the safe handling and transport of containers. It decrees that every container travelling internationally be fitted with a CSC Safety-approval Plate. This holds essential information about the container, including age, registration number, dimensions and weights, as well as its strength and maximum stacking capability.

Longshoremen and related unions around the world struggled with this revolution in shipping goods. For example, by 1971 a clause in the International Longshoremen's Association (ILA) contract stipulated that the work of "stuffing" (filling) or "stripping" (emptying) a container within 50 miles of a port must be done by ILA workers or if not done by ILA that the shipper needed to pay royalties and penalties to the ILA. Unions for truckers and consolidators argued that the ILA rules were not valid work preservation clauses because the work of stuffing and stripping containers away from the pier had not traditionally been done by ILA members. In 1980 the Supreme Court of the United States heard this case and ruled against the ILA.

Description

Forty foot (12.2 m) containers make up 70% of the world's container volume, which is measured in TEUs

Ninety percent of the global container fleet consists of *"dry freight"* or *"general purpose"* containers – both of standard and special sizes. And although lengths of containers vary from 8 to 56 feet (2.4 to 17.1 m), according to two 2012 container census reports about 80% of the world's containers are either twenty or forty foot standard length boxes of the dry freight design. These typical containers are rectangular, closed box models, with doors fitted at one end, and made of corrugated weathering steel (commonly known as CorTen) with a plywood floor. Although corrugating the sheet metal used for the sides and roof contributes significantly to the container's rigidity and stacking strength, just like in corrugated iron or in cardboard boxes, the corrugated sides cause aerodynamic drag, and up to 10% fuel economy loss in road or rail transport, compared to smooth-sided vans.

Standard containers are 8-foot (2.44 m) wide by 8 ft 6 in (2.59 m) high, although the taller "High Cube" or "hi-cube" units measuring 9 feet 6 inches (2.90 m) have become very common in recent years. By the end of 2013, high-cube 40 ft containers represented almost 50% of the world's maritime container fleet, according to Drewry's Container Census report.

About 90% of the world's containers are either nominal 20-foot (6.1 m) or 40-foot (12.2 m) long, although the United States and Canada also use longer units of 45 ft (13.7 m), 48 ft (14.6 m) and 53 ft (16.15 m). ISO containers have castings with openings for twistlock fasteners at each of the eight corners, to allow gripping the box from above, below, or the side, and they can be stacked up to ten units high. Regional intermodal containers, such as European and U.S. domestic units however, are mainly transported by road and rail, and can frequently only be stacked up to three laden units high. Although the two ends are quite rigid, containers flex somewhat during transport.

The standard castings on the eight corners of each container. The twistlock proper is done through a larger oval hole on the top or bottom.

Container capacity is often expressed in twenty-foot equivalent units (TEU, or sometimes *teu*). A twenty-foot equivalent unit is a measure of containerized cargo capacity equal to one standard 20-foot (6.1 m) long container. This is an approximate measure, wherein the height of the box is not considered. For example, the 9 ft 6 in (2.9 m) tall high-cube, as well as 4-foot-3-inch (1.3 m) half-height 20-foot (6.1 m) containers are equally counted as one TEU. Similarly, extra long 45 ft (13.72 m) containers are commonly designated as two TEU, no different than standard 40 feet (12.19 m) long units. Two TEU are equivalent to one forty-foot equivalent unit (FEU).

In 2014 the global container fleet grew to a volume of 36.6 million TEU, based on Drewry Shipping Consultants' Container Census. Moreover, in 2014 for the first time in history 40-foot High cube containers accounted for the majority of boxes in service, measured in TEU.

Manufacturing prices for regular, dry freight containers are typically in the range of $1750—$2000 U.S. per CEU (container equivalent unit), and about 90% of the world's containers are made in China. The average age of the global container fleet was a little over 5 years from end 1994 to end 2009, meaning containers remain in shipping use for well over 10 years.

Bottom Structure Features

Forty foot or longer containers typically have a *gooseneck tunnel*, an indentation in the floor structure, that meshes with the gooseneck on dedicated container semi-trailers. The gooseneck tunnel is clearly visible in the underside of a toppled-over container (first picture), as well as in a container's interior, where it takes the space otherwise covered by wood flooring. Gooseneck container trailer showing twistlock couplings for forty-foot boxes at its four corners. Twenty foot containers on the other hand, frequently have forklift pockets, accessible from the sides (last picture).

Types

Other than the standard, general purpose container, many variations exist for use with different cargoes. The most prominent of these are refrigerated containers (a.k.a. *Reefers*) for perishable goods, that make up six percent of the world's shipping boxes. And tanks in a frame, for bulk liquids, account for another 0.75% of the global container fleet.

Forty foot High-cube actively refrigerated container – refrigerating equipment visible on the front end.

Although these variations are not of the standard *type*, they mostly are *ISO standard* containers – in fact the ISO 6346 standard classifies a broad spectrum of container types in great detail. Aside from different *size* options, the most important container types are:

- General-purpose dry vans, for boxes, cartons, cases, sacks, bales, pallets, drums, etc., Special interior layouts are known, such as:
    - rolling-floor containers, for difficult-to-handle cargo
    - garmentainers, *for shipping garments on hangers (GOH)*
- Ventilated containers. Essentially dry vans, but either passively or actively ventilated. For instance for organic products requiring ventilation
- Temperature controlled – either insulated, refrigerated, and/or heated containers, for perishable goods
- Tank containers, for liquids or gases. Frequently these are dangerous goods, and in the case of gases one shipping unit may contain multiple gas bottles
- Bulk containers (sometimes *bulktainers*), either closed models with roof-lids, or hard or soft open-top units for top loading, for instance for bulk minerals. Containerized coal carriers and "bin-liners" (containers designed for the efficient road and rail transportation of rubbish from cities to recycling and dump sites) are used in Europe.

A spine car with a 20 ft tank container and an open-top 20 ft container with canvas cover

- Open-top and open-side containers, for instance for easy loading of heavy machinery or oversize pallets. Open sides are also used for ventilating hardy perishables like apples or potatoes.
- Platform based containers such as:
    - flat-rack and bolster containers, for barrels, drums, crates, and any heavy or bulky out-of-gauge cargo, like machinery, semi-finished goods or processed timber. Empty flat-racks can either be stacked or shipped sideways in another ISO container
    - collapsible containers, ranging from flushfolding flat-racks to fully closed ISO and CSC certified units with roof and walls when erected.

Containers for Offshore use have a few different features, like pad eyes, and must meet additional strength and design requirements, standards and certification, such as the DNV2.7-1 by Det Norske Veritas and the European standard EN12079: *Offshore Containers and Associated Lifting Sets.*

A flat-rack container loaded with a small vessel loaded by a reach stacker.

A multitude of equipment, such as generators, has been installed in containers of dif-ferent types to simplify logistics.

Swap body units usually have the same bottom corner fixtures as intermodal containers, and often have folding legs under their frame so that they can be moved between trucks without using a crane. However they frequently don't have the upper corner fittings of ISO containers, and are not stackable, nor can they be lifted and handled by the usual equipment like reach-stackers or straddle-carriers. They are generally more expensive to procure.

Specifications

Basic dimensions and permissible gross weights of intermodal containers are largely determined by two ISO standards:

Two 45-foot 'High-cube' containers on a Roll-on/roll-off (RoRo) tractor. The 9 ft 6 height of the boxes is identified by diagonal yellow and black markings on the top corners of the container

- ISO 668:2013 Series 1 freight containers—Classification, dimensions and ratings
- ISO 1496-1:2013 Series 1 freight containers—Specification and testing—Part 1: General cargo containers for general purposes

Load bearing of container stacking is at the 40 ft coupling

Weights and dimensions of the most common *standardized* types of containers are given below. Values vary slightly from manufacturer to manufacturer, but must stay within the tolerances dictated by the standards. Empty weight (*tare weight*) is not determined by the standards, but by the container's construction, and is therefore indicative, but necessary to calculate a net load figure, by subtracting it from the maximum permitted gross weight.

imperial		20□ container		40□ container		40□ high-cube container		45□ high-cube container	
		metric	imperial	metric	imperial	metric	imperial	metric	
external dimensions	length	19′ 10.5″	6.058 m	40′ 0″	12.192 m	40′ 0″	12.192 m	45′ 0″	13.716 m
	width	8′ 0″	2.438 m	8′ 0″	2.438 m	8′ 0″	2.438 m	8′ 0″	2.438 m
	height	8′ 6″	2.591 m	8′ 6″	2.591 m	9′ 6″	2.896 m	9′ 6″	2.896 m
interior dimensions	length	19′ 3″	5.867 m	39′ 5 45/64″	12.032 m	39′ 4″	12.000 m	44′ 4″	13.556 m
	width	7′ 8 19/32″	2.352 m	7′ 8 19/32″	2.352 m	7′ 7″	2.311 m	7′ 8 19/32″	2.352 m
	height	7′ 9 57/64″	2.385 m	7′ 9 57/64″	2.385 m	8′ 9″	2.650 m	8′ 9 15/16″	2.698 m
door aperture	width	7′ 8 ⅛″	2.343 m	7′ 8 ⅛″	2.343 m	7′ 6″	2.280 m	7′ 8 ⅛″	2.343 m
	height	7′ 5 ¾″	2.280 m	7′ 5 ¾″	2.280 m	8′ 5″	2.560 m	8′ 5 49/64″	2.585 m
internal volume		1,169 ft^3	33.1 m^3	2,385 ft^3	67.5 m^3	2,660 ft^3	75.3 m^3	3,040 ft^3	86.1 m^3

maximum gross weight	66,139 lb	30,400 kg	66,139 lb	30,400 kg	68,008 lb	30,848 kg	66,139 lb	30,400 kg
empty weight	4,850 lb	2,200 kg	8,380 lb	3,800 kg	8,598 lb	3,900 kg	10,580 lb	4,800 kg
net load	61,289 lb	28,200 kg	57,759 lb	26,200 kg	58,598 lb	26,580 kg	55,559 lb	25,600 kg

Stacking Containers

53' 48' 45' 40' and (2x) 20' containers stacked

At stacking load-bearing locations, 40-foot containers are the standard unit length, and 45 ft, 48 ft, and 53 ft all stack at the 40 ft coupling width. Other units can be stacked on top of 20 ft units only if there are two in a row (40 ft coupling width) but 20 ft units can not be stacked on top of 40 ft units, or any other larger container.

Forty-five-foot containers can be seen sticking out 2½ feet, as part of the forty foot container stacks at the back of this ship.

The coupling holes are it takes a double male twist lock to securely mate stacked containers together.

Non-standard and Uncommon Sizes

Pallet Wide Containers

Pallet Wide containers have about 4 inches (10.2 cm) more internal floor width than standard containers to accommodate more Euro-pallets, common in Europe. These containers typically have an internal width of 2.44 m ($96\frac{1}{8}$ in), to be able to load either two or three of the 1.2 m ($47\frac{1}{4}$ in) long by 0.8 m ($31\frac{1}{2}$ in) wide pallets side by side. Many sea shipping providers in Europe allow these as overhangs on standard containers are sufficient and they fit in the usual interlock spaces (or with the same floor panel the side ribs of pallet-wide containers are embossed to the outside instead of being molded to the inside).

Especially the 45 ft (13.72 m) pallet-wide high-cube shortsea container has gained wider acceptance, as these containers can replace the 13.6 m (44 ft $7\frac{3}{8}$ in) swap bodies that are common for truck transport in Europe. The EU has started a standardization for pallet wide containerization in the European Intermodal Loading Unit (EILU) initiative.

Australian RACE containers are also slightly wider to optimise them for the use of Australia Standard Pallets.

48-foot Containers

The 48-foot (14.63 m) shipping container is a High Cube container in that it is 9 ft 6 in (2.90 m) tall on the exterior. It is 8 ft 6 in (2.59 m) wide which makes it 6 inches (15 cm) wider than ISO-standard containers. This size was introduced by container shipping company APL in 1986, and is used domestically in North America on road and rail, and may be transported on deck by ship. This size being 8 feet (2.44 m) longer and 6 inches (15 cm) wider has 29% more cubic capacity than the standard 40-ft High Cube, yet the cost to move it by truck or rail are almost the same.

53-foot Containers

Swift 53 ft Intermodal container

General purpose 53-foot (16.15 m) containers were introduced in the United States in 1989, and are used both in the U.S.A. and Canada, mainly for domestic road and rail transport. They are considered High-cubes, based on their 9 ft 6 in (2.90 m) ISO-standard height. Their width of 8 ft 6 in (2.59 m) however makes them 6 inches (15 cm) wider than ISO-standard containers. These large boxes have 60% more capacity than standard-height 40-foot (12.19 m) containers, enabling shippers to consolidate more cargo into fewer containers.

Generally, North American 53-foot containers were not constructed strongly enough to endure the rigors of ocean transport, but in 2007 container carrier APL introduced the first 53-foot ocean-capable containers. All new, reinforced 53-foot boxes were built specifically for international trade and designed to withstand ocean voyages on its South-China to Los Angeles service. In 2013 however, APL stopped offering vessel space for 53-foot containers on its trans-Pacific ships. Nevertheless, In 2015 both Crowley and TOTE Maritime each announced the construction of their respective second combined container & RoRo ships for Puerto Rico trade, with the specific design to maximize cubic cargo capacity by carrying 53-foot, 102-inch (2,591 mm)-wide containers. Within Canada, Oceanex offers 53-foot-container ocean service to and from the island of Newfoundland. Fifty-three-foot containers are also being used on some Asia Pacific international shipping routes.

Small Containers

The United States military continues to use small containers, strongly reminiscent of their Transporter and Conex boxes of the 1950s and 1960s. These either comply with ISO standard dimensions, or are a direct derivative thereof. Current terminology of the US armed forces calls these small containers Bicon, Tricon and Quadcon, which correspond with ISO 668 standard sizes 1D, 1E and 1F respectively. This comes down to containers of 8 ft (2.44 m) height, and with a footprint size either one half (Bicon), one third (Tricon) or one quarter (Quadcon) the size of a standard 20-foot, one TEU container. At a nominal length of 10 feet (3.0 m), two Bicons coupled together *lengthwise* match one 20-foot ISO container, but their height is 6 inches (15 cm) shy of the more commonly available 10-foot ISO containers of so-called *standard* height, which are 8 ft 6 in (2.59 m) tall. Tricons and Quadcons however have to be coupled *transversely* – either three or four in a row – to be stackable with twenty foot containers. Their *length* of 8 ft (2.44 m) corresponds to the *width* of a standard 20-foot container, which is why there are forklift pockets at their ends, as well as in the sides of these boxes, and the doors only have one locking bar each. The smallest of these, the Quadcon, exists in two heights: 96 in (2.44 m) or 82 in (2.08 m). Only the first conforms to ISO-668 standard dimensions (size 1F).

Reporting Mark

Various markings on the rear end of a container

Each container is allocated a standardized ISO 6346 reporting mark (ownership code), four letters long ending in either U, J or Z, followed by six digits and a check digit. The ownership code for intermodal containers is issued by the *Bureau International des Containers* (International container bureau, abbr. B.I.C.) in France, hence the name BIC-Code for the intermodal container reporting mark. So far there exist only four-letter BIC-Codes ending in "U".

The placement and registration of BIC Codes is standardized by the commissions TC104 and TC122 in the JTC1 of the ISO which are dominated by shipping companies. Shipping containers are labelled with a series of identification codes that includes the manufacturer code, the ownership code, usage classification code, UN placard for hazardous goods and reference codes for additional transport control and security.

Following the extended usage of pallet-wide containers in Europe the EU had started the Intermodal Loading Unit (ILU) initiative. This showed advantages for intermodal transport of containers and swap bodies. This led to the introduction of ILU-Codes defined by the standard EN 13044 which has the same format as the earlier BIC-Codes. The International Container Office BIC agreed to only issue ownership codes ending with U, J or Z. The new allocation office of the UIRR (International Union of Combined Road-Rail Transport Companies) agreed to only issue ownership reporting marks for swap bodies ending with A, B, C, D or K – companies having a BIC-Code ending with U can allocate an ILU-Code ending with K having the same preceding letters. Since July 2011 the new ILU codes can be registered, beginning with July 2014 all intermodal ISO containers and intermodal swap bodies must have an ownership code and by July 2019 all of them must bear a standard-conforming placard.

Handling

A cargo container being transferred from a rail car to a flat-bed truck, lifted by a reach stacker

Containers are transferred between rail, truck, and ship by container cranes at container terminals. Forklifts, reach stackers, straddle carriers, and cranes may be used to load and unload trucks or trains outside of container terminals. Swap bodys, sidelifters, tilt deck trucks, and hook trucks allow transfer to and from trucks with no extra equipment.

ISO-standard containers can be handled and lifted in a variety of ways by their corner fixtures, but the structure and strength of 45-foot (type E) containers limits their tolerance of side-lifting, nor can they be forklifted, based on ISO 3874 (1997).

Transport

Containers can be transported by container ship, truck and freight trains as part of a single journey without unpacking. Units can be secured in transit using "twistlock" points located at each corner of the container. Every container has a unique BIC code painted on the outside for identification and tracking, and is capable of carrying up to 20–25 metric tons. Costs for transport are calculated in twenty-foot equivalent units (TEU).

Rail

When carried by rail, containers may be carried on flatcars or well cars. The latter are specially designed for container transport, and can accommodate double-stacked containers. However the loading gauge of a rail system may restrict the modes and types of container shipment. The smaller loading gauges often found in European railroads will only accommodate single-stacked containers. In some countries, such as the United Kingdom, there are sections of the rail network through which high-cube containers cannot pass, or can pass through only on well cars. On the other hand, Indian Railways runs double-stacked containers on flatcars under 25 kV overhead electrical wires. The wires must be at least 7.45 metres (24 ft 5 in) above the track, but IR is able to do so because of its large loading gauge and the extra stability provided by its 1,676 mm (5 ft 6 in) track. China Railways also runs double-stacked containers under overhead wires, but must use well cars to do so, since the wires are only 6.6 metres (21 ft 8 in) above the track and 1,435 mm (4 ft 8 $^1/_{2\,in}$) (standard gauge) does not provide adequate stability to run double-stacked containers on flat cars.

A portion of a "double stack" container train operated by Union Pacific Railroad, the containers are owned by Pacer Stacktrain, the well cars by DTTX.

Ship

About 90% of non-bulk cargo worldwide is transported by container, and the largest container ships can carry over 19,000 TEU. Each year an estimated 10,000 shipping containers fall into the sea; of these 10% are expected to contain chemicals toxic to marine life.

Plane

Containers can also be transported in planes, as seen within intermodal freight transport. However, transporting containers in this way is typically avoided due to the cost of doing such and the lack of availability of planes which can accommodate such awkwardly sized cargo.

Securing and Security

Securing Containers and Contents

There are many established methods and materials for stabilizing and securing intermodal containers loaded on ships, as well as the internal cargo inside the boxes. Conventional restraint methods and materials such as steel strapping and wood blocking and bracing have been around for decades and are still widely used. Polyester strapping and lashing, and synthetic webbings are also common today. Dunnage bags (also known as "air bags") are used to keep unit loads in place.

Flexi-bags can also be directly loaded, stacked in food-grade containers. Indeed, their standard shape fills the entire ground surface of a 20' ISO container.

Dockworkers securing containers on a ship with steel lashing bars and turnbuckles	Containers can be horizontally connected with lashing bridge fittings	Polyester Lashing Application	Polyester Strapping and Dunnage Bag application

Security

Intermodal containers which contain valuables can be the target of break-ins and burglary when left unattended. In these cases, the container may be fitted with a security system consisting of a motion detector and panel inside the container. The panel can trigger a siren, strobe, or light to deter intruders, or use a radio signal to alert security guards.

Items that were packed incorrectly may come loose and cause a false response from an inside motion detector. If criminals break in by cutting through a wall of the container, the obstructed motion detector becomes useless. Tomographic motion detectors work well in intermodal containers because they do not require a line of sight to detect motion. The entire container is covered by a volumetric sensing mesh that is not blocked by equipment or inventory. Tomographic motion detection is not prone to misdetection due to dirt buildup as is the case for beams and infrared sensors.

Non-shipping Uses

Containerized Equipment

Container-sized units are also often used for moving large pieces of equipment to temporary sites. Specialised containers are particularly attractive to militaries already using containerisation to move much of their freight around. Shipment of specialized equipment in this way simplifies logistics and may prevent identification of high value equipment by enemies. Such systems may include command and control facilities, mobile operating theatres or even missile launchers (such as the Russian 3M-54 Klub surface-to-surface missile).

File:Hammelmann Diesel unit – built into container

Complete water treatment systems can be installed in containers and shipped around the world.

Electric generators can be permanently installed in containers to be used for portable power.

Repurposing

Container City in Cholula, Mexico uses fifty old sea containers for 4,500 m² (48,000 sq ft) of workshops, restaurants, galleries, etc., as well as some homes.

Containers have long been used for other purposes, typically but not always at the end of their voyaging lives. US military often used their Conex containers as on-site storage, or easily transportable housing for command staff and medical clinics. Nearly all of over 150,000 Conex containers shipped to Vietnam remained in country, primarily as storage or other mobile facilities. Permanent or semi-permanent placement of containers for storage is common. A regular forty-foot container has about 4,000 kg (8,818 lb) of steel, which takes 8,000 kWh (28,800 MJ) of energy to melt down. Repurposing used shipping containers is increasingly a practical solution to both social and ecological problems.

Shipping container architecture employs used shipping containers as the main framing of modular home designs, where the steel may be an integrated part of the design, or be camouflaged into a traditional looking home. They have also been used to make temporary shops, cafes, and computer datacenters, *e.g.*, the Sun Modular Datacenter.

Intermodal containers are not strong enough for conversion to underground bunkers without additional bracing, as the walls cannot sustain much lateral pressure and will collapse. Also, the wooden floor of many used containers could contain some fumigation residues, rendering them unsuitable as confined spaces, such as for prison cells or bunkers. Cleaning or replacing the wood floor can make these used containers habitable, with proper attention to such essential issues as ventilation and insulation.

Pallet

The classic wooden pallet

A plastic pallet with nine legs, which can be lifted from all four sides

A metal pallet with removable beams, in this case for firewood

Automated palletizer of bread with industrial KUKA robots at a bakery in Germany

48" × 40" galvanized steel pallet. Galvanized steel pallets are fireproof and rust resistant.

A pallet, sometimes inaccurately called a skid (a skid has no bottom deck boards), is a flat transport structure that supports goods in a stable fashion while being lifted by a forklift, pallet jack, front loader, work saver, or other jacking device, or a crane. A pallet is the structural foundation of a unit load which allows handling and storage efficiencies. Goods or shipping containers are often placed on a pallet secured with strapping, stretch wrap or shrink wrap and shipped. Since its invention in the twentieth century, its use has dramatically supplanted older forms of crating like the wooden box and the wooden barrel, as it works well with modern packaging like cardboard boxes and Intermodal containers commonly used for bulk shipping.

While most pallets are wooden, pallets can also be made of plastic, metal, paper, and recycled materials. Each material has advantages and disadvantages relative to the others.

Overview

Containerization for transport has spurred the use of pallets because the shipping containers have the smooth, level surfaces needed for easy pallet movement. Most pallets can easily carry a load of 1,000 kg (2,205 lb). Today, over half a billion pallets are made each year and about two billion pallets are in use across the United States alone.

Pallets make it easier to move heavy stacks. Loads with pallets under them can be hauled by forklift trucks of different sizes, or even by hand-pumped and hand-drawn pallet jacks. Movement is easy on a wide, strong, flat floor: concrete is excellent. The greatest investment needed for economical pallet use is in the construction of commercial or industrial buildings. Passage through doors and buildings must be possible. To help this issue, some later pallet standards (the europallet and the U.S. Military 35 in × 45.5 in or 889 mm × 1,156 mm) are designed to pass through standard doorways.

Organizations using standard pallets for loading and unloading can have much lower costs for handling and storage, with faster material movement than businesses that do not. The exceptions are establishments that move small items such as jewelry or large items such as cars. But even they can be improved. For instance, the distributors of costume jewelry normally use pallets in their warehouses and car manufacturers use pallets to move components and spare parts.

The lack of a single international standard for pallets causes substantial continuing expense in international trade. A single standard is difficult because of the wide variety of needs a standard pallet would have to satisfy: passing doorways, fitting in standard containers, and bringing low labor costs. For example, organizations already handling large pallets often see no reason to pay the higher handling cost of using smaller pallets that can fit through doors.

Due to cost and a need to focus on core business, pallet pooling becomes more and more common. Some pallet suppliers supply users with reusable pallets, sometimes with integral tracking devices. A pallet management company can help supply, clean, repair, and reuse pallets.

Pallets should be seen as reusable packaging items as a pallet can potentially be used multiple times over the course its lifetime. In the UK, government legislation relating to the Waste Framework Directive requires the reuse of packaging items above recycling and disposal.

A truckload of used wooden pallets in California

Standardization and Regulation

Dimensions

Wooden pallets typically consist of three or four stringers that support several deckboards, on top of which goods are placed. In a pallet measurement the first number is the stringer length and the second is the deckboard length. Square or nearly square pallets help a load resist tipping.

Two-way pallets are designed to be lifted by the deckboards. The standard North American pallet, or GMA pallet, has deckboards of 40 inches and stringers of 48 inches.

Four-way pallets, or pallets for heavy loads (or general-purpose systems that might have heavy loads) are best lifted by their more rigid stringers. These pallets are usually heavier, bigger and more durable than two-way pallets.

Pallet users want pallets to easily pass through buildings, to stack and fit in racks, to be accessible to forklifts and pallet jacks and to function in automated warehouses. To avoid shipping air, pallets should also pack tightly inside intermodal containers and vans.

No universally accepted standards for pallet dimensions exist. Companies and organizations utilize hundreds of different pallet sizes around the globe. While no single dimensional standard governs pallet production, a few different sizes are widely used.

ISO Pallets

The International Organization for Standardization (ISO) sanctions six pallet dimensions, detailed in *ISO Standard 6780: Flat pallets for intercontinental materials handling—Principal dimensions and tolerances*:

Dimensions (W × L) millimetres	Dimensions (W × L) inches	Wasted floor, ISO container	Region most used in
1016 × 1219	40.00 × 48.00	3.7% (20 pallets in 40 ft ISO)	North America
1000 × 1200	39.37 × 47.24	6.7%	Europe, Asia; similar to 40" × 48".
1165 × 1165	45.9 × 45.9	8.1%	Australia
1067 × 1067	42.00 × 42.00	11.5%	North America, Europe, Asia
1100 × 1100	43.30 × 43.30	14%	Asia
800 × 1200	31.50 × 47.24	15.2%	Europe; fits many doorways

North American Pallets

Of the top pallets used in North America, the most commonly used by far is the Grocery Manufacturers' Association (GMA) pallet, which accounts for 30% of all new wood pallets produced in the United States. The ISO also recognizes the GMA pallet footprint as one of its six standard sizes.

Dimensions, mm (W × L)	Dimensions, in (W × L)	Production Rank	Industries Using
1016 × 1219	40 × 48	1	Grocery, many others
1067 ×1067	42 × 42	2	Telecommunications, Paint
1219 × 1219	48 × 48	3	Drums
1219 × 1016	48 × 40	4	Military, Cement
1219 × 1067	48 × 42	5	Chemical, Beverage
1016 × 1016	40 × 40	6	Dairy
1219 × 1143	48 × 45	7	Automotive
1118 × 1118	44 × 44	8	Drums, Chemical
914 × 914	36 × 36	9	Beverage
1219 × 914	48 × 36	10	Beverage, Shingles, Packaged Paper
889 × 1156	35 × 45.5	Unknown	Military ½ ISO container, fits 36" standard doors
1219 × 508	48 × 20	Unknown	Retail

European Pallets

EURO pallet type	Dimensions (W × L)		ISO pallet alternative
EUR, EUR 1	800 mm × 1,200 mm	31.50 in × 47.24 in	ISO1, same size as EUR
EUR 2	1,200 mm × 1,000 mm	47.24 in × 39.37 in	ISO2
EUR 3	1,000 mm × 1,200 mm	39.37 in × 47.24 in	
EUR 6	800 mm × 600 mm	31.50 in × 23.62 in	ISO0, half the size of EUR

	600 mm × 400 mm	23.62 in × 15.75 in	quarter the size of EUR
	400 mm × 300 mm	15.75 in × 11.81 in	one-eighth the size of EUR

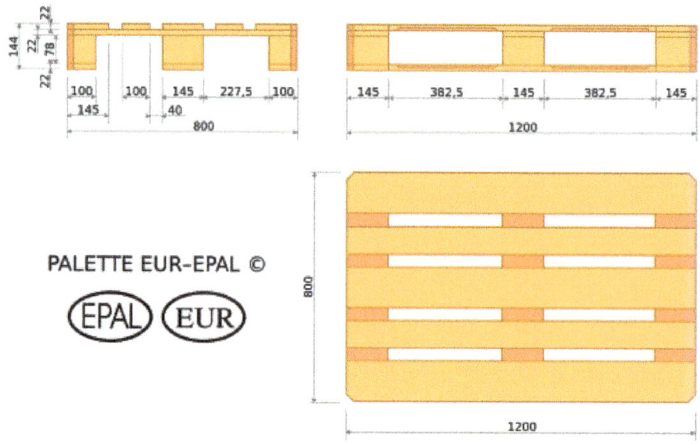

EuroPallet

Australian Standard Pallets

The Australian standard pallet is a pallet size commonly found in Australia but found rarely elsewhere. It is a square pallet originally made of hardwood 1,165 mm × 1,165 mm (45.87 in × 45.87 in) in size which fits perfectly in the RACE container of the Australian Railway. They are ill-suited for the standard 20 feet (6.1 m) and 40 feet (12 m) ISO shipping containers used around the globe. Australian standard pallets are usually manufactured in hardwood, but 1165 × 1165 mm pallets can also be manufactured using lighter timber suitable for use as disposable pallets using 16 millimetre boards. Extensively used in storage and warehousing, they are popular pallets for racking, with the right shape and size to be removed from transport and directly onto warehouse racking for storage.

The Australian Standard Pallet dates back to World War II, while ISO containers date to the late 1950s. Although the pallet's dimensions pre-date the ISO containers, it requires less dunnage, is square, and leaves less wasted space than other pallets, including the GMA pallet. In 2010, Australia adopted the globally accepted ISPM 15 wood packaging material regulations (before this time it was hardwood and more expensive).

Standard-setting Organizations

A number of different organizations and associations around the world work towards establishing and promulgating standards for pallets. Some strive to develop universal standards for pallet dimensions, types of material used in construction, performance standards, and testing procedures. Other organizations choose to focus on pallet standards for a specific industry (such as groceries) or type of material (such as wood).

ISO Technical Committee 51: Pallets for Unit Load Method of Materials Handling

ISO TC 51 states its scope of work entailing the "standardization of pallets in general use in the form of platforms or trays on which goods may be packed to form unit loads for handling by mechanical devices". The Technical Committee works in conjunction with other Technical Committees focused on transportation infrastructure to develop interrelated standards. TC 51 is responsible for developing *ISO Standard 6780: Flat pallets for intercontinental materials handling—Principal dimensions and tolerances* as well as sixteen other standards related to pallet construction and testing.

National Wood Pallet and Container Association

The National Wood Pallet and Container Association (NWPCA) is a trade organization based in the United States representing the interests of wood pallet and container manufacturers.

U.S. DOD, Dept. of Navy, Naval Sea Systems Command

This organization maintains MIL-STD-1660, the standard description of palletized unit loads for the U.S. Military and some allies.

DOD Unit loads generally use 40 in × 48 in (1,016 mm × 1,219 mm) pallets, are less than 4,000 lb (1,814 kg), weatherproof, and stack 16 ft (4.88 m) high. They often use steel pallets, steel straps with notched seals, outdoor plywood, and plastic film. The standard describes tests for stacking, transport, sling, forklift and pallet jack, impact, drop tests, tip, water-retention, and disassembly.

European Committee for Standardization (Comité Européen De Normalisation)

In addition to the other standards it publishes, the European Committee for Standardization, also known as the Comité Européen de Normalisation (CEN), produces standards for pallets. While the standards are voluntary in nature, many companies and organizations involved in transportation have adopted them. The major standard for pallets produced by CEN is *ICS: 55.180.20 General purpose pallets*

Phytosanitary Compliance

Due to the International Plant Protection Convention (abbreviated IPPC), most pallets shipped across national borders must be made of materials that are incapable of being a carrier of invasive species of insects and plant diseases. The standards for these pallets are specified in ISPM 15.

Pallets made of non-wood materials such as steel, aluminum, plastic, or engineered wood products, such as plywood, oriented strand board, or corrugated fiberboard do not need IPPC approval, and are considered to be exempt from ISPM 15 regulations.

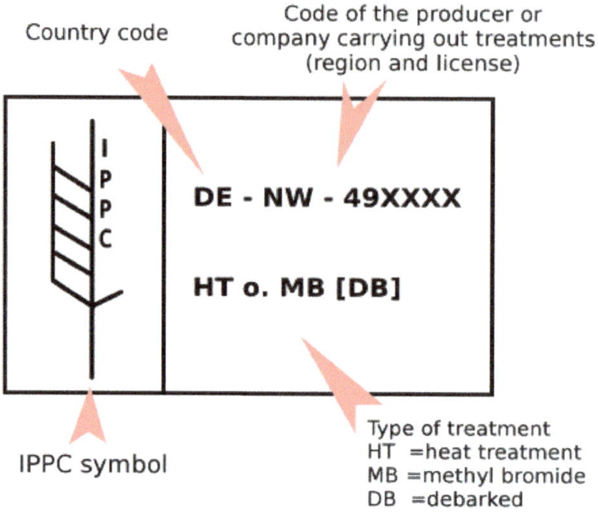

IPPC marks on a pallet from Germany (DE)

Pallets made of raw, untreated wood are not compliant with ISPM 15. To be compliant the pallets (or other wood packaging material) must meet debarked standards, and must be treated by either of the following means under the supervision of an approved agency:

- Heat treatment The wood must be heated to achieve a minimum core temperature of 56 °C (132.8 °F) for at least 30 minutes. Pallets treated via this method bear the initials HT near the IPPC logo.
- Chemical fumigation The wood must be fumigated with methyl bromide. Pallets treated via this method bear the initials MB near the IPPC logo. From 19 March 2010 the use of Methyl Bromide as an acceptable treatment according to ISPM15 has now been banned within all EU member states. This is due to causing potential harm to the Earth's stratospheric ozone layer.

Treated wood pallets must be stamped on two opposite sides indicating either HT for heat treated or MB for methyl bromide treatment.

Pallet Construction

Types

Although pallets come in all manner of sizes and configurations, all pallets fall into two very broad categories: "stringer" pallets and "block" pallets. Various software packages exist to assist the pallet maker in designing an appropriate pallet for a specific load, and to evaluate wood options to reduce costs.

Pallets being used in a warehouse in Finland

Stringer Pallet

Stringer pallets use a frame of three or more parallel pieces of timber (called stringers). The top deckboards are then affixed to the stringers to create the pallet structure. Stringer pallets can have a notch cut into them allowing "four-way" entry. Forklifts can lift a stringer pallet from all four directions, though lifting by the stringers is more secure.

Block Pallet

Block pallets are typically stronger than stringer pallets. Block pallets utilize both parallel and perpendicular stringers to better facilitate efficient handling. A block pallet is also known as a "four-way" pallet, since a pallet-jack may be used from any side to move it.

Perimeter Base Pallet

All stringer and some block pallets have "unidirectional bases," i.e. bottom boards oriented in one direction. While automated handling equipment can be designed for this, often it can operate faster and more effectively if the bottom edges of a pallet have bottom boards oriented in both directions. For example, it may not need to turn a pallet to rack it, and operation is less sensitive to pallet orientation.

Quality Improvements

The least expensive way to improve a pallet is usually to specify better nails. With non-wood pallets, a controlled coefficient of friction is often helpful to prevent the pallet from slipping from forks and racks. Stiffer pallets are more durable, and are handled more easily by automated equipment. If a pallet does not need to be lifted from all four sides, two-way pallets with unnotched stringers may be used, with the additional benefits of added rigidity and strength. Specifying tolerances on flatness and water content may help the supplier meet target requirements. Inspection of pallets, whether in person or by a third-party (such as "SPEQ" inspected pallets) offer additional assurance of quality.

Materials

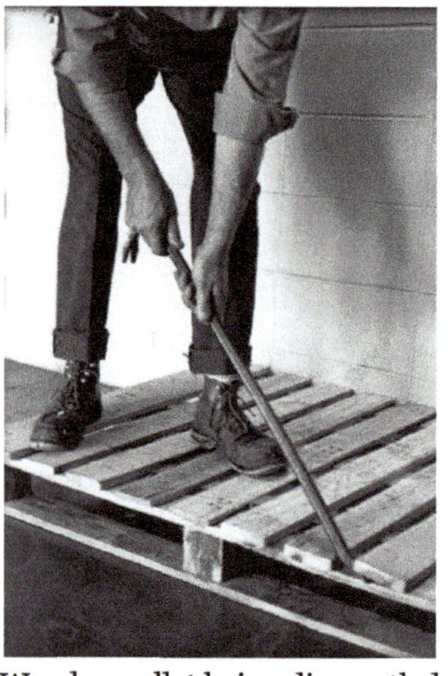

Wooden pallet being dismantled

The cheapest pallets are made of softwood and are often considered expendable, to be discarded as trash along with other wrapping elements, at the end of the trip. These pallets are simple stringer pallets, and liftable from two sides.

Slightly more complex hardwood block pallets, plastic pallets and metal pallets can be lifted from all four sides. These costlier pallets usually require a deposit and are returned to the sender or resold as used. Many "four way" pallets are color-coded according to the loads they can bear, and other attributes. Wood pallets can pose serious bio-hazard risks as they are susceptible to bacterial and chemical contamination, such as *E. coli* problems in food and produce transportation, and even insect infestation, and thus the need for ISPM 15.

Wooden pallet construction specifications can depend on the pallet's intended use: general, FDA, storage, chemical, export; the expected load weight; type of wood desired: recycled, hard, soft, kiln dried or combo (new & recycle); and even the type of fasteners desired to hold the pallet together: staples or nails.

Paper pallets, also referred to as "ecopallets", are often used for light loads, but engineered paper pallets are increasingly used for loads that compare with wood. Paper pallets are also used where recycling and easy disposal is important. New designs of ecopallets have been made from just two flat pieces of corrugated board (no glue/staples) and weigh just 4.5 kg (9.9 lb), offering dramatic freight savings. Ecopallets are also ISPM 15 exempt, negating fumigation and barrier "slip" sheets. They are cleaner, safer, and provide a cost-saving eco-friendly alternative to other pallet materials. Some engineered Corrugated pallets offer a significant reduced height, providing substantial freight cost reduction. Low profile hand pallet trucks allow picking up pallets as low as 25 mm.

Steel pallets are strong and are used for heavy loads, high-stacking loads, long term dry storage, and loads moved by abusive logistic systems. They are often used for military ammunition. Metal pallets make up less than 1% of the market. Materials include carbon steel, stainless steel, and aluminum. Of these, carbon steel offers excellent durability at the lowest cost. Stainless steel doesn't require a paint coating, and is preferred for such applications as clean room environments. Carbon steel units are expensive compared to wood, and stainless and aluminum cost about 2–3 times that of carbon steel. Long term costs, however, can be lower than wood. General advantages of metal pallets are high strength and stiffness, excellent durability, bug free, no splinters, sanitary, and recyclable. Disadvantages include a higher initial price, significant weight, low friction, and susceptibility to rusting (carbon steel). Metal is primarily used in captive or closed loop environments where durability and product protection are key performance requirements. Metal units today are increasingly price competitive and lighter in weight. Primary industries that use metal pallets include automotive, pharmaceutical, lawn tractors, motorcycles, and tires.

Aluminum pallets are stronger than wood or plastic, lighter than steel, and resist weather, rotting, plastic creep and corrosion. They are sometimes used for air-freight, long-term outdoor or at-sea storage, or military transport.

Plastic

Plastic pallets are often made of new HDPE or recycled PET (drink bottles). They are usually extremely durable, lasting for a hundred trips or more, and resist weathering, rot, chemicals and corrosion. The benefits of plastic pallets over wood pallets include the ability to be easily sanitized, resistance to odor, fire retardancy, longer service life span, durability and better product protection, non-splintering, and lighter weight, thus saving on transportation and labor costs and making them safer and more environmentally friendly.They often stack. Plastic pallets are exempt by inspection for biosafety concerns, and easily sanitize for international shipping. HDPE is impervious to most acids and toxic chemicals clean from them more easily. Some plastic pallets can collapse from plastic creep if used to store heavy loads for long periods. Plastic pallets cannot easily be repaired, and can be ten times as expensive as hardwood, so they are often used by logistics service providers who can profit from their durability and stackability. The large supply chains have increased the use of plastic pallets as many organisations seek to reduce costs through waste, transport and health & safety. Pallets and dolly combined for example the Pally eliminate pallet instability and the need for additional lifting equipment, but also create valuable space in busy operating environments and deliver significant time and cost savings by reducing supply chain handling. Plastics' reusability has contributed to an increase in usage of plastic pallets.

Plastic pallets are produced and used widely in the U.S. and Europe, spurred by the adoption of the ISPM 15. A full comparison of wood vs plastic can be made by a life cycle analysis. Plastic pallets can cost 10 times as much as hardwood pallets and even more expensive compared to cheap expendable softwood pallets.

RFID chips can be molded into the pallets to monitor locations and track inventory.

Manufacture

There are six main types of plastic processes that are used to manufacture pallets:

- High pressure injection molding
- Structural foam molding
- Thermoforming
- Compression molding
- Rotational molding
- Profile extrusion

History

Skids go back to Ancient Egypt and Ancient Mesopotamia at least as far back as the 1st millennium B.C.

The development of the forklift and the needs of World War II logistics operations led to substantial use of pallets.

References to the early modern pallets types are slim with a string of patents showing parts of the development. The earliest may be a U.S. patent on a skid from 1924 describing Hallowell's "Lift Truck Platform." In the late 1930s, pallets became more commonplace with the newer forklift types. George Raymond filed for a patent in 1938 (granted US Patent 2178646 in 1939) for a pallet designed to complement a new pallet jack design; the essential features of both are still in common use today. A 1939 patent from Carl Clark shows type of pallet with steel stringers. War time developments were often just patented after the war, so there is a patent from Robert Braun on a four-way pallet in 1945, and a patent from Norman Cahners (a U.S. Navy Supply Officer) shows a disposable pallet type in 1949. The complete construction principle of a modern four-way pallet is described by Darling Graeme in 1949.

Alternative Uses

Old and discarded wooden pallets can be used in pallet crafts and various furniture pieces.

Stacked pallets

Discarded wooden pallets should not be used for firewood or crafts unless it has been determined that the wood in these pallets has not been treated with wood preservatives, fungicides and/or pesticides. Various pyrethrins and propiconazole are common treatments for wooden pallets. In addition, imported palletized goods are routinely fumigated with highly toxic pesticides. During use, harmful materials or chemicals also may spill on the pallet wood and be absorbed.

Improvised street bench in Naples (Italy), made from wooden EuroPallet

Craft publications have advised readers to use pallets to build a skateboarding obstacle called a manual pad, barricades during amateur paintball games, or other sport-related items. Other publications have suggested using pallet wood for small animal cages or fences. Pallet wood has been recycled for use as furniture wood by at least one company.

The acoustic guitar maker Taylor Guitars once produced a high quality "pallet guitar" made from pallet wood, in order to demonstrate the importance of construction technique versus expensive exotic woods.

I-Beam Design, an architecture and interior design firm based in New York, NY won an award in a 1999 competition sponsored by Architecture for Humanity for their submission of "The Pallet House", a design solution to house the returning refugees of Kosovo. Full-scale prototypes of The Pallet House were featured in the "Casa per Tutti" Exhibit at the Milan Triennale and the Earth Awards in HRH Prince Charles' Royal Gardens as part of The Prince's Charities Foundation's Conference on a Sustainable Future organized in collaboration with IBM and The Financial Times. The Pallet House is an affordable transitional home that can become permanent over time. It can be used as refugee housing or as affordable housing as well. Due to the nature of the Pallet module, the wall cavity can be insulated with a variety of materials that are accessible to the user.

The two Austrian students *Andreas Claus Schnetzer* and *Gregor Pils* from the University of Vienna created a home entitled *Pallet house* and as the name suggests, reused pallets to form a modular, energy efficient and affordable housing. The idea was born in 2008 during a competition and the Pallet house has been exhibited in several European cities including Venice, Vienna, Linz and Grenoble. It could become a clever approach to low income housing.

Denver, Colorado was host to an Inaugural Pallet-Fest festival that showcased the versatility of pallets and included large art structures made with upcycled materials, a pallet maze, a pallet amphitheater with live musical performances, sustainable living demonstrations, an upcycled fashion show, vendors and artists selling upcycled goods, and a parkour course. The event was crowdfunded and organized by Upcycle Events.

Items made from pallet wood are likely to be durable and demonstrate good weather resistance due to these treatments. However, close contact with pallet wood or inhalation of dusts from sanding or sawing can be a source of exposure to pesticide and fungicide chemicals. It is likely that the January, 2010 recall of Johnson and Johnson Tylenol and other drugs was due to their being stored on wooden pallets that had been treated with the fungicide/pesticide 2,4,6-tribromophenol. This chemical can be degraded by molds to produce 2,4,6-tribromoanisole whose strong, musty odor caused consumers to complain. There is no acute or chronic health data on 2,4,6-tribromoanisole, but it is believed that the contaminated drugs caused nausea and other health effects in some people.

Fire Hazards

Both wood and plastic pallets are possible fire hazards. The National Fire Protection Association requires that both types "shall be stored outside or in a detached structure" unless protected by fire sprinklers.

Food Safety Risks

Wood pallets used to transport food can possibly harbor pathogens such as *E. coli* and Listeria.

A release by the National Consumers League announced the testing of 70 wood pallets and 70 plastic pallets that have been loaded with perishable products and shipped to an end user was shipped overnight to an independent microbiology lab for testing. The results came back with 10 percent of the wood pallets positive for *E. coli* and 1.4 percent positive on the plastic pallets.

Pallet Racking

Pallet racking made from Dexion slotted angle

Pallet rack is a material handling storage aid system designed to store materials on pallets (or "skids"). Although there are many varieties of pallet racking, all types allow for the storage of palletized materials in horizontal rows with multiple levels. Forklift trucks are usually an integral part of any pallet rack system as they are usually required to place the loaded pallets onto the racks for storage. Since the Second World War, pallet racks have become an essential and ubiquitous element of most modern warehouses, manufacturing facilities, retail centers, and other storage and distribution facilities.

Push back system designed to store pallets of motorcycles

All types of pallet racking increase storage density of the stored goods. Costs associated with the racking increases with increasing storage density.

Pallet flow lane showing rollers

Selective Pallet Racking Systems

Selective pallet racking is a common pallet racking system in use today. Selective pallet racking systems typically come in one of two configurations: a roll formed, or clip-in configuration, and a structural bolt-together configuration.

1. Roll formed selective rack is most commonly manufactured in a "teardrop" style (so named as the holes on the column of the upright are shaped like a teardrop). Pallets then rest on the horizontal load beams that are held in place by mounting clips. Because the clips on teardrop configurations can be quickly moved, the shelves can be easily adjusted to different heights to accommodate various load sizes. This is convenient for a warehouse that needs to store a wide variety of product sizes.

2. Structural pallet rack systems are very similar to roll formed pallet rack systems except the horizontal load beams are attached to the uprights with bolts and have much greater weight-bearing capacity. The bolt fixings make this a form of adjustable shelving - racks can be constructed, reconfigured, and dismantled and reused as necessary.

Structural pallet racking can be designed into the structure of the building itself, so that the upright columns are simultaneously used to support the roof of the storage facility, in which case the structural pallet rack uprights replace the storage building's vertical support I-beams. This system is a rack supported building.

Selective pallet rack systems provide easy accessibility to all products at all times - important if the inventory is rapidly depleted and restocked (called quick turnover). A selective pallet rack system is commonly used in a "big-box" distribution application, as well as in retail store inventory rooms, cold storage applications, wholesale stores, etc.

Common components of selective rack include the following:

1. Load beams (also called step beams or box beams).

- Most step beams are roll formed members with a 1⅝" step along the top inside edge. This step is used to hold any load support components such as pallet supports or wire decks.

- Box beams have no inset step, instead have four flat sides like a box. All load beams typically mount onto an upright frame column with integral rivets or hooks. Some systems utilize an extra clip or bolt to lock the beam to the upright.

- Structural beams are hot formed structural C shapes with connecting clips at either end. Structural load beams are generally used with structural uprights but can be used with standard roll formed uprights. Structural load beams offer heavier weight capacities than step beams or box beams.

2. Upright frames (also called upright columns or uprights) vary in size and design depending on load requirements, and styles. The most common upright column is produced by roll forming flat coil stock steel into a modified "C" shape with returns. This style is often referred to as open-back roll-formed columns. Holes or slots are punched during manufacturing up and down the column at standard intervals so that the load beams can be mounted into the upright columns. Upright frames can also be constructed utilizing structural C shapes for columns. Structural uprights have an increased weight capacity over roll formed uprights.

3. Diagonal braces and horizontal braces, commonly referred to as upright frame lacing is usually welded between two upright columns to form selective upright frames. The lacing may be bolted to the columns in some cases.

4. Pallet supports are roll formed channels that are placed front to back between the load beams to support pallets.

5. Timber decking comes in two forms: close boarded and slated. Close boarded decking provides a solid 'floor' to a level of racking while slatted is formed by rows of timber and acts as a permanent pallet.

6. Wire decking is commonly used as a safety measure on selective pallet rack to prevent pallets or the products stored on them from falling through the rack structure. Wire mesh decking comes in various thicknesses and mesh dimensions. Wire mesh construction also allows for easy identification of shelf contents and prevents dirt and other debris from accumulating on the shelves because of the holes in the mesh. Most wire mesh decking has U-shaped channel supports, also known as struts, to support the load. With this waterfall decking, the wire mesh extends across the top and down the front of the beam to provide more support, and is more desirable in the marketplace. Reverse water-

fall decking can provide containment of a loose product to prevent the product from falling behind the rack system. Lay-in decking rests inside the step of the beam, and wire mesh does not waterfall over the beam. Some types of decking are manufactured with solid metal instead of wire mesh. Even though the solid decking provides a greater distributed weight capacity, it is discouraged by fire inspectors because sprinkler systems cannot spray through the shelves to levels below.

7. Footplates, also known as footpads or baseplates, are at the base of columns and serve as anchors to give the rack more stability: anchor bolts are inserted through the baseplate's holes to attach the column to the concrete floor. Footplates are made of thicker steel and in some geographic locations, they must be of a certain size and seismic rating. Footpads increase the pallet rack's overall stability and weight-bearing capacity.

8. Shims are used when the uprights are resting on uneven floors; the shims, equal in size to the base of the uprights, are installed beneath the uprights to level the rack.

9. Row spacers are sometimes used if uprights are arranged in back-to-back rows; the spacers are mounted between adjacent columns to ensure that the rows are kept straight and to give the pallet racks even more strength and steadiness.

10. Wall ties may be used for further support if the uprights are arranged in a row along a wall.

11. Column protectors, also known as post protectors, are protective shields that can be installed around the base of an upright to minimize damage where forklifts might hit the upright. Damage to the base of a column can weaken the entire frame and could cause it to collapse. Column protectors are made of various materials such as polyethylene, ductile iron casting, and other durable materials.

12. Guard rails are installed to increase protection for upright columns and for human safety when platforms or steps are attached to pallet racks.

Very narrow aisle otherwise known as VNA, is the use of selective pallet racking in a tighter configuration to provide maximum space utilization within a storage facility. These systems typically operate in conjunction with wire-guided or rail-guided reach-truck systems. A wire-guided system consists of a wire embedded in the concrete floor that provides tracking for the reach-truck. A rail-guided system consists of angle iron bolted to the floor down the length of each row. Typically, the angle iron is 4" by 3" and ¼" - ⅜" inches thick. A distinct advantage of a narrow aisle pallet racking is fast picking without large aisles which results in improved use of space. When there is limited space, a compact storage method is ideal. Fully adjustable system flexibility and space saving aisles can be manipulated as one to give the greatest amount of pallet storage locations.

Other Common Types

Many types of pallet storage racks are available with different designs to fulfill specific functions or create specific advantages. To create the ideal pallet racking system, several considerations should be taken into account:

- Desired storage density
- Floor space and building height
- Placement of building doors and columns
- Inventory accessibility
- Inventory rotation
- Item/load size and weight
- Optimal storage design
- Cost of materials and installation

Some of the Most Common Types of Pallet Rack Systems Used Include:

1. Drive-in and drive-through (sometimes spelled Drive-thru) are storage rack configurations that allow the forklift to drive directly into the lane of stacked rows (called a bay). The difference between a drive-in and a drive-thru pallet rack system is simply whether the bays have an entry at only one end, or at both ends. Drive-in rack systems use a common entry and exit, while drive-thru systems have entry points at either end of the bay. Because a drive-in racking system has only one entrance, it uses what is called a LIFO (last in, first out) storage method. With only one entrance, the last pallet put into a row is necessarily the first one to be taken out. A drive-thru storage system, with two different entry points, can also use a FIFO (first in, first out) storage method. With a FIFO system, pallets are loaded in one end and are pushed back to the other end, where they are then at the front of the row on the opposite side. The first pallet put into such a row is the first one taken out at the other end. This system is advantageous for material with an expiration date or wherever shelf life is a major concern.

2. Push-back pallet rack systems are designed around the principle of organizing space by depth rather than width. This depth arrangement greatly reduces aisle

space and increases storage density. In this configuration, each bay can be up to six pallets deep; each pallet stored on wheeled carts that fit onto rails. The rails are slightly angled toward the load/unload side of the rack in order to take advantage of gravity, saving enormous amounts of energy for moving heavy pallets. When a forklift sets the pallet onto the cart, it drives forward and causes the pallet to bump the next pallet, causing the entire row of pallets to roll backwards. When removing a pallet from the front position the remaining pallets immediately stage themselves forward so that the next available pallet can be accessed. Push back rack is a LIFO (last in, first out) storage system.

3. Pallet flow systems are high density pallet storage systems that utilize depth to increase capacity. This system uses a slightly inclined rail with rollers that allow pallets to move easily along the sloped plane. These systems are also called gravity flow

or dynamic flow systems. The pallet flow system often has complex motion and braking systems to control the speed of the moving pallet. Pallet flow racking systems are either a FIFO (first in, first out) or a LIFO (last in, first out) storage system. If the system is loaded from the back and unloaded from the front, it is FIFO; if the system is loaded and unloaded from the front it is a LIFO system.

1. Compact mobilized pallet racks also known as motorized mobile pallet racks or mobile industrial rack system, are designed to maximize the use of warehouse storage space. Mobilized compact pallet racks can reduce your pallet storage area in half, or double the capacity of your pallet rack storage in the same floor-space area. Mobilized pallet racks save space by converting static access aisles into productive storage space. The mobilized compact rack storage system mounts rows of pallet racks onto heavy-duty rolling carriages and places them on floor tracks. Rows of pallet racks (up to 96' long) move sideways along the floor tracks creating a movable floating access aisle. Typical warehouse storage facilities may have dozens of static access aisles but only one or two forklifts capable of accessing stored materials. Mobilizing pallet racks reduce the number of access aisles to match the velocity requirements of your storage area. Rows of pallet racks are moved by pushing a button or by remote control, which opens access aisles when and where they are needed.

Mobilized pallet racks have helped many organizations reduce or eliminate new building construction costs, auxiliary warehouses, building expansions, and decrease ongoing operational costs (lighting, energy, insurance) by maximizing the use of their storage floor space.

Some disadvantages of high density pallet storage systems are; less access to all stock at any given moment (although if the stored product is all the same, it should not matter), and the expense of such systems. Selective pallet rack systems are considerably less expensive per pallet position than their higher density counterparts. In most medium to large facilities, however, high density pallet rack systems are essential, since they provide the efficiency of time and high cost facility space is better optimized.

Identifying the brand of racking systems is important to order compatible shelving or resell the racks. It is possible to differentiate between the many pallet rack brands by looking at the shape of the vertical beams, increments of height adjustment and the shape of the holes on the vertical beams.

Safety Considerations

Because of the size and weight of pallets, important safety factors have to be considered at all times:

1. Pay attention to any loose components in the pallet rack system, and take the time to report any damage in the pallet rack frame; such frame damage could cause the pallets to fall.
2. It is the owner's legal responsibility to communicate this important warning to all who are around storage racks:
   "Never climb on racks during or after assembly. Storage racks are not designed to be stepped on or climbed on. A slip or fall may result in serious injury."
   It is especially important to have highly visible warning signs if the pallet rack system is used in retail environments, such as wholesale centers, where the public is present.
3. Use only quality pallets that are not damaged. To save money, or perhaps from neglectful management, some warehouses use pallets until they become faulty and dangerous. Regular inspection of pallets for broken or fractured planks or stringers, protruding nails, and missing support blocks is essential. Damaged pallets can cause loading and unloading problems; for example, loose stringers can get hung up on the pallet racks, which can cause loads to fall from high positions. Also, faulty pallets can cause obstruction problems in flow systems by jamming certain pallet rack designs.
4. Always ensure that the proper motorized equipment is being used for the application.
5. Do not obstruct the end of aisles by staging pallets in these areas. Doing so can cause severe and potentially fatal injuries and accidents.
6. Never overload or exceed the recommended load specifications for a racking system. Overloading may cause a catastrophic failure of your storage rack system.

7. Rack audits (safety checks) should be performed on a regular basis by a qualified inspector familiar with RMI design and safety standards to make sure all aspects of system are up to 100% operationally safety.

8. Food products must follow FDA regulations and use approved pallet racks. Your location may also require different equipment configurations.

Bulk Box

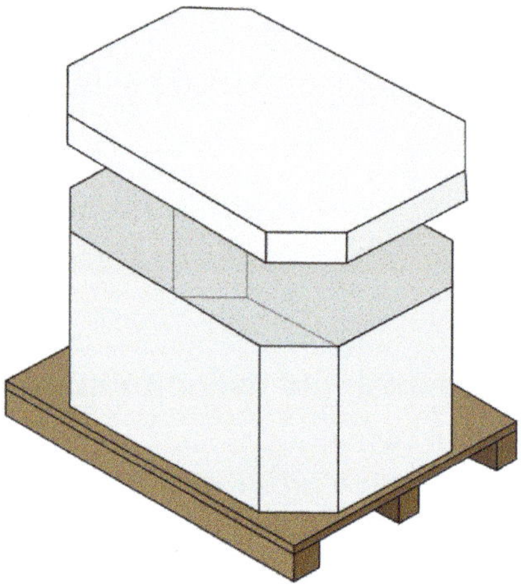

Octagonal corrugated bulk box (octabin), with cover

A bulk box, also known as a bulk bin, skid box, pallet box, bin box, or octabin is a pallet-size box used for storage and shipping of bulk quantities.

Bulk boxes are often made of corrugated fiberboard, either doublewall or triplewall. Many corrugated bulk boxes have covers. The main body of some is similar to a *half slotted container* with flaps on the bottom; others have a separate base (similar to the cover) and a sleeve for side walls. Additional corrugated liners and box reinforcement are sometimes used to control bulging. Wooden boxes are also used for bulk packaging; boxes made of aluminum and steel are common in heavy industry. Reusable plastic totes (molded or corrugated plastic) are used for some products and logistics chains.

For some products, inner plastic liners or bin bags are used to protect the contents. Sealed inner bags with integral valves are sometimes used for liquids. Most bulk boxes are recyclable, returnable, or reusable. Some styles of boxes fold (knock down) or come apart for easier return shipments.

In the U.S. and Canada, the term *gaylord* is sometimes used for triplewall corrugated pallet boxes; this is due to the first bulk bins being manufactured by the original Gaylord Container Company of St. Louis, which was acquired by Crown Zellerbach of San Francisco in 1955.

Use

Humanitarian aid shipped in pallet boxes. Cover secured with strapping

Reusable steel bins or racks for bulk products

Bulk boxes are used for loose parts, mixed small containers, granular materials, powders, liquids, etc. Use in industry is common: shipping and storage of bulk intermediate materials prior to further processing or packaging. They are sometimes used to collect and ship materials for recycling and waste.

When used to ship hazardous waste or dangerous goods, box construction and use are highly regulated.

ASTM Standards

- D5168 Standard Practice for Fabrication and Closure of Triple-wall Corrugated Containers
- D6179 Standard Test Methods for Rough Handling of Unitized Loads and Large Shipping Cases and Crates
- D6251 Standard Specification for Wood-Cleated Panelboard Shipping Boxes
- D6254 Standard Specification for Wirebound Pallet-Type Wood Boxes
- D6256 Standard Specification for Wood-Cleated Shipping Boxes and Skidded, Load-Bearing Bases
- D6573 Standard Specification for General Purpose Wirebound Shipping Boxes
- D6880-05 Standard Specification for wooden boxes

Crate

A crate is a large shipping container, often made of wood, typically used to transport or store large, heavy or awkward items. Steel and aluminium crates are also used. Specialized crates were designed for specific products, and were often made to be reusable, such as the "bottle crates" for milk and softdrinks.

Crates can be made of wood, plastic, metal or other materials. The term 'crate' often implies a large and strong container. Most plastic crates are smaller and are more commonly called a case or container. Metal is rarely used because of its weight. When metal is used, a crate is often constructed as an 'open crate' and may be termed a 'cage'. Although a crate may be made of any material, for these reasons, the term 'crate' used alone often implies one constructed of wood.

Wooden Crates

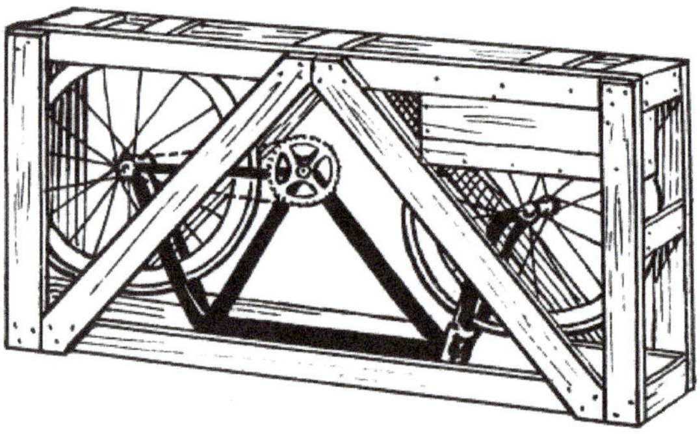

An open frame style crate used for a bicycle

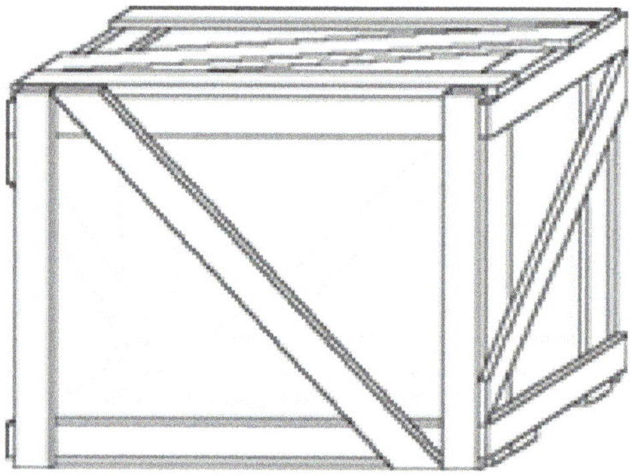

An open style crate

A wooden crate has a self-supporting structure, with or without sheathing. For a wooden container to be a crate, all six of its sides must be put in place to result in the rated strength of the container. Crates are distinct from wooden boxes. The strength of a wooden box is rated based on the weight it can carry before the top (top, ends, and sides) is installed, whereas the strength of a crate is rated with the top in place. In general conversation, the term crate is sometimes used to denote a wooden box.

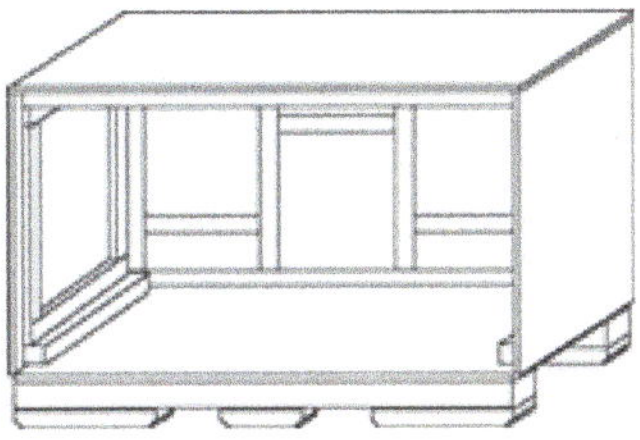	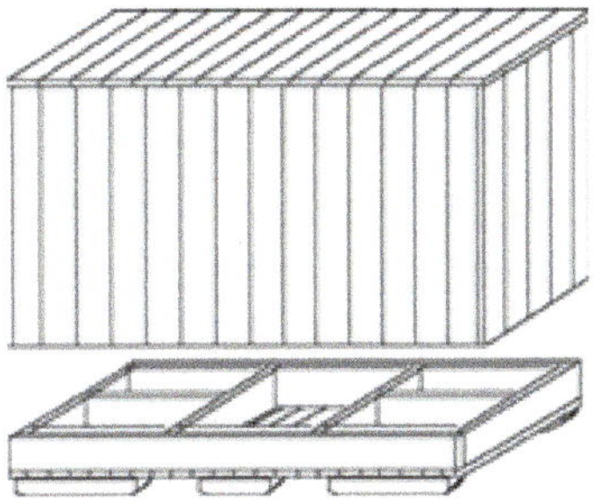	
A sheathed (closed) plywood crate	A sheathed (closed) lumber crate	Wooden crate with cleating

History

The first documented reference to a shipping crate in the United States is in a 1930 handbook, *Technical Bulletin No. 171* written by C. A. Plaskett for the U.S. Department of Agriculture. His writing implies that crates were defined before that time. Plaskett was known for his extensive testing and defining of various components of transport packaging. The USDA Forest Service revised and expanded it in 1964 as the "Wood Crate Design Manual", Handbook 252.

Construction

Although the definition of a wooden crate, as compared to a wooden box, is clear, construction of the two often results in a container that is not clearly a crate or a box. Both wooden crates and wooden boxes are constructed to contain unique items, the design of either a crate or box may use principles from both. In this case, the container will typically be defined by how the edges and corners of the container are constructed. If the sheathing (either plywood or lumber) can be removed, and a framed structure will remain standing, the container would likely be termed a crate. If removal of the sheathing results in no way of fastening the lumber around the edges of the container, the container would likely be termed a wooden box.

Design

There are many variations of wooden crate designs. By far the most common are 'closed', 'open' and 'framed'. A Closed Crate is one that is completely or nearly completely enclosed with material such as plywood or lumber boards. When lumber is used, gaps are often left between the boards to allow for expansion. An Open Crate is one that (typically) uses lumber for sheathing. The sheathing is typically gapped by at various distances. There is no strict definition of an open crate as compared to a closed crate. Typically when the gap between boards is greater than the distance required for expansion, the crate would be considered an open crate. The gap between boards would typically not be greater than the width of the sheathing boards. When the gap is larger, the boards are often considered 'cleats' rather than sheathing thus rendering the crate unsheathed. An unsheathed crate is a frame crate. A Frame Crate is one that only contains a skeletal structure and no material is added for surface or pilferage protection. Typically an open crate will be constructed of 12 pieces of lumber, each along an outer edge of the content and more lumber placed diagonally to avoid distortion from torque.

When any type of crate reaches a certain size, more boards may be added. These boards are often called Cleats. A cleat is used to provide support to a panel when that panel has reached a size that is may require added support based on the method of transportation. Cleats may be placed anywhere between the edges of a given panel. On crates, cleat placement is often determined by the width of the plywood used on plywood sheathed crates. On other crates, cleats are often evenly spaced as required to strengthen the panel. Sometimes two cleats are added across the top panel of a crate placed as needed to give the top of the crate added strength where lifting chains or straps may press on the crate while lifting.

Cleats may have more specific names based on added benefit they provide. Some published standards only use those more descriptive terms and may never refer to these various lumber components as cleats. For example, lumber placed under the top of a wood container to add support for a large top are called "joists". Lumber is built into the midsection of the top of a wood container to strengthen the top are called "cleats". When the cleats are enlarged and constructed to support a large top, they may generically be termed "cleats" or more specifically be termed "joists".

"Skids" or thick bottom runners, are sometimes specified to allow forklift trucks access for lifting.

Transportation methods and storage conditions must always be considered when designing a crate. Every step of the transportation chain will result in different stresses from shock and vibration. Differences in pressure, temperature and humidity may not only adversely affect the content of the crate, but also will have an effect on the holding strength of the fasteners (mostly the nails and staples) in the crate. In some countries, any wooden crate being designed to ship overseas must be treated to ISPM 15 standards or commonly known as the "bug stamp" to prevent the spread of disease and insects.

Although the above definition almost always stands true, there are many slightly altered or 'sub-definitions' used by and in various organizations, agencies and documents. This is the result of the small size of the industry and the fact that a single, finite definition of an item that is different every time it is made can be difficult to define.

IATA, the International Air Transport Association, for example, doesn't allow crates on airplanes because it defines a crate as an open transport container. Although a crate can be of the Open or Framed variety, having no sheathing, a Closed crate is not open and is equally as safe to ship in as a wooden box, which is allowed by IATA.

A plywood sheathed crate with a pallet-like bottom.

Distribution Center Management System

Distribution Center Management System (DCMS) is a user friendly Warehouse Management System (WMS), designed to track the activities performed in a Distribution Center (DC)/ Warehouse. It is created and owned by a private company called Eclipse Systems Pvt Ltd. It automates the entire process flow of receiving, managing and shipping goods to customers from the warehouse. DCMS solutions are designed for both large and small scale businesses. In January 2015, the product went open source.

DCMS helps warehouses to remove the risk of stock pile-up, stock outs, pending orders and loss of sales due to customer dissatisfaction. It dramatically improves the warehouse productivity and increase inventory accuracy to 99.8%.

It helps strengthening customer relationships, reduce operating expenses and increase warehouse and distribution efficiencies. Due to its modular design, DCMS can be easily customized for various clients as per their requirements. WMS is an intelligent investment that will instantly boost accuracy in the warehouse operations and yields long term financial benefits.

DCMS Modules

A typical list of modules for DCMS can be as follows:

- receiving
- putaway
- order processing
- replenishment
- pulling

- restocking
- picking
- validation
- sorting
- shipping

Features of a DCMS also has the following:

- crate management
- supervisory control
- reports

Benefits

An effective DCMS helps to prevent stock outs, stock pile-ups, pending orders and loss of sales due to customer dissatisfaction. It improves the warehouse productivity and increases inventory accuracy too. Some other benefits include reducing operating expenses and improvement in efficiencies.

Radio Frequency and Paper Based Solution

DCMS works in both Radio Frequency (RF) as well as paper based environment. DCMS is a future ready solution. Warehouses which do not wish to make heavy investments in RF immediately, have the option of choosing paper based environment first, and can later adopt RF technology with minimal time and effort.

Third Party Integration

DCMS provides an open interface and can be easily integrated with third party software applications. It reduces unnecessary paper work and office routines. Having DCMS in place means warehouses do not need to depend on someone's experience any more.

DCMS vs Warehouse Module of Enterprise Resource Planning

Inventory initiative in recent times such as JIT (Just In Time), ECR (Efficient Customer Response) and QR (Quick Response) have turned traditional inventory management users to WMS application users. Not only this, more and more Enterprise Resource Planning (ERP) users are replacing their existing warehouse module of ERP with WMS because of its functionality fit, knowledgebase, latest up gradations, integration, adaptation to ever changing customer requirement which the ERP warehouse module cannot provide.

System Architecture

DCMS is based on Microsoft Technology with an Oracle as Database. It uses web based three tier architecture that provides a standard suite of warehousing applications. DCMS can also be installed on cloud and used as SAAS. DCMS provides rich source of WEB based reports that can be used by all levels of organizations for making important decisions.

Future Ready Solution

Think of a warehouse problem, DCMS will have comprehensive solution for that. Not only this, it is also ready for all future possibilities for which the warehouse may not be prepared in the current scenario. Eclipse System's rich knowledge base and R&D department keeps the application up to date to provide the latest industry features.

Drop Shipping

Drop shipping is a supply chain management method in which the retailer does not keep goods in stock but instead transfers customer orders and shipment details to either the manufacturer, another retailer, or a wholesaler, who then ships the goods directly to the customer. As in retail businesses, the majority of retailers make their profit on the difference between the wholesale and retail price, but some retailers earn an agreed percentage of the sales in commission, paid by the wholesaler to the retailer.

Procedure

To enable customers to review items similar to those they can purchase, a drop shipping retailer might keep "display items" on display in a physical "brick and mortar" store, or provide a product catalog as either hard copy or online, or even some combination of all these methods ("bricks and clicks").

Retailers that drop ship merchandise from wholesalers can take measures to hide this fact or keep the wholesale source from becoming widely known. This can be achieved by "blind shipping" (shipping merchandise without a return address), or "private label shipping" (having merchandise shipped from the wholesaler with a return address customized to the retailer). The wholesaler might include a customized packing slip, including details such as the retailer's company name, logo, and contact information.

Drop shipping can occur when a small retailer (that typically sells in small quantities to the general public) receives a single large order for a product. Rather than route the shipment through the retail store, the retailer might arrange for the goods to be shipped directly to the customer. Drop shipping is also very common with big ticket items like steel buildings, where the retailer will take a deposit and have the steel shipped directly to the buyer's building site from the supplier's manufacturing facility.

Many sellers, including those on online auction sites, also drop ship. Often, a seller will list an item as new and ship the item directly from the retailer or wholesaler to the customer or highest bidder. The seller profits from the difference between the sales of the product or winning bid and the wholesale price, less any selling and merchant fees, or shipping fees. On eBay there is not a clear rule about drop shipping. According to their policy, this practice should be avoided, as sellers should always own the products before publishing them. This is to prevent bad buyer experiences and to guarantee safe transactions. However, it seems that eBay is tolerating drop shipping as long as the seller can guarantee good service.

One effect of drop shipping is that customers who receive a drop-shipped package will realize that they overpaid for the item on eBay, return the item to the manufacturer, then reorder the identical item directly from the manufacturer. The cost of processing the return and the loss of the unsellable returned product can result in significant losses to the manufacturer.

An emerging trend in the drop ship business is private label drop shipping, in which a manufacturer produces a custom item for a retailer and drop ships it. The range of private label drop shipped items available includes, for example, simple keepsakes, apparel with custom logos, pictures, and customized formulations for vitamins and nutritional supplements.

Developments

Major drop shipping suppliers and fulfillment services are primarily based in the USA.

Since 2006 many drop ship companies have emerged in China, many of which offer wholesale and drop shipping services to both companies and individuals. This is largely due to the increasing ease of e-procurement and the growing part that the internet is playing in e-commerce. Drop ship suppliers based in China have increasingly been able to compete with same-country distributors because of improved logistics for small packets and the easing of trade barriers.

Benefits

Two significant benefits of drop shipping are the elimination of upfront inventory and a positive cash-flow cycle. A positive cash flow cycle occurs because the seller is paid when the purchase is made. The seller usually pays the wholesaler using a credit card or credit terms. Therefore, there is a period of time in which the seller has the customer's money, but has not yet paid the wholesaler.

Drop shipment can also be an advantage in some transactions since it (can) obscure the actual shipper from the final receiver, thus preserving a middleman's role by preventing direct communication/awareness between the original seller and final buyer, without requiring a middleman to bear the costs/delay in physically taking in the goods, repackaging, relabeling, etc.

Drop shipping also eliminates some duplication of effort since only one warehouse will pick, pack, and ship the product. This approach can reduce total inventory management and shipping costs. These cost reductions can subsequently reduce the price to the consumer.

Risks

Some risks such as backordering may occur when a seller places a shipment request with a wholesaler, but the product is sold out. Backordering may be accompanied by a long wait for a shipment while the wholesaler waits for new products, which may reflect badly on the retailer. A good wholesaler will keep retailers updated, but it is the business owner's job to be aware of the quantities that the wholesaler has available.

Drop shipping has also featured prominently in some Internet-based home business scams. Scam artists will promote drop shipping as a lucrative "work from home opportunity". The victim who buys into this scam will be sold a list of businesses from which drop shipment orders can be placed. These businesses may not be wholesalers, but other businesses or individuals acting as middlemen between retailers and wholesalers, with no product of their own to sell. These middlemen often charge prices that leave little profit margin for the victim, and require a regular fee for the retailer's usage of their services.

Related Concepts

The converse of drop shipping is *will call,* where a customer picks up directly from a wholesaler rather than via a retail shop.

As with drop shipping, affiliate marketing allows a blogger, website owner, or other form of internet content owners commonly known in that industry as publishers or partners, to list or promote a product or advertisement campaign on behalf of a third party brand typically called merchants. In affiliate marketing the product or advertisements are not owned or designed by the affiliate publisher and often the affiliates and their content must adhere to contractual standards detailing how the brand and promotion may be represented. Affiliates often may not imply in their marketing that they own or represent the brand. In contrast with drop shipping, the potential customer, often called a lead, is typically redirected to the third party merchant's own website, sales portal,or shopping cart. Thus in traditional affiliate marketing transactions the lead is redirected to the merchant's own sales campaign and so are only processed directly through the third party's network. Therefore, an affiliate commission is generated for leads and sales occurring directly on the third party merchant's website and would not be combined with any processes or commissions arising from a drop shipping relationship.

References

- Levinson, Marc (2006). "7". The Box: How the Shipping Container Made the World Smaller and the World Economy Bigger (PDF). Princeton, N.J.: Princeton University Press. p. 127. ISBN 0-691-12324-1. Retrieved 21 July 2015.

- Van Ham, Hans; Rijsenbrij, Joan (15 December 2012). Development of Containerization. Amsterdam: IOS Press. p. 26. ISBN 978-1614991465. Retrieved 14 August 2015.
- Soroka, W (2008). Illustrated Glossary of Packaging Terms. Institute of Packaging Professionals. p. 33. ISBN 1-930268-27-0.
- Raballand, Gaël; Aldaz-Carroll, Enrique (1 February 2005). "How Do Differing Standards Increase Trade Costs? The Case of Pallets". Retrieved 1 September 2016.
- Andeerson, L O; Heebink, T B (1964), Wood Crate Design Manual (PDF), Handbook 252, US Forest Products Laboratory, USDA, retrieved 10 June 2015
- Grande, Joseph A. (September 2008). "Plastic Pallets Gain Ground In an Eco-Conscious World". Plastics Technology. Retrieved 8 June 2014.
- Singh, R. R. (2013-07-18). "Pallets – An efficient and safe way of material handling" (PDF). Bureau of Indian Standards. Retrieved 2014-01-22.
- ISO 1496-1: Series 1 Freight Containers – Specification and Testing (PDF) (Technical report). ISO. 1990. pp. 8, 13, 20. Part 1. Archived from the original (PDF) on 29 March 2014.
- "40ft High Cubes set to Dominate the Container Equipment Market". Drewry Shipping Consultants. 18 June 2014. Archived from the original on 29 August 2014. Retrieved 18 July 2015.
- "Composition of the Global Fleet of Containers, 2008". The Geography of Transport Systems. Hofstra University. Archived from the original on 21 November 2014. Retrieved 18 July 2015.
- Michael J. Everhart (7 July 2014). "My Vietnam Tour – 1970". Retrieved 21 July 2015. .. CONEX ... container that ... was about 7' high by 8' wide and about 12' long...
- Lewandowski, Krzysztof (2014). "Czechoslovak activity to prepare European norms for containers before the Second World War" (PDF). Acta Logistica. 1 (4): 1–7. ISSN 1339-5629.
- "Container Shipping – Statistics & Facts". Statista.com. Statista Inc. Archived from the original on 2014-12-05. Retrieved 27 July 2015.
- "World Container Fleet Overview". csiu.co. CSI Container Services International. January 2014. Archived from the original on 18 July 2015. Retrieved 18 July 2015.

Permissions

All chapters in this book are published with permission under the Creative Commons Attribution Share Alike License or equivalent. Every chapter published in this book has been scrutinized by our experts. Their significance has been extensively debated. The topics covered herein carry significant information for a comprehensive understanding. They may even be implemented as practical applications or may be referred to as a beginning point for further studies.

We would like to thank the editorial team for lending their expertise to make the book truly unique. They have played a crucial role in the development of this book. Without their invaluable contributions this book wouldn't have been possible. They have made vital efforts to compile up to date information on the varied aspects of this subject to make this book a valuable addition to the collection of many professionals and students.

This book was conceptualized with the vision of imparting up-to-date and integrated information in this field. To ensure the same, a matchless editorial board was set up. Every individual on the board went through rigorous rounds of assessment to prove their worth. After which they invested a large part of their time researching and compiling the most relevant data for our readers.

The editorial board has been involved in producing this book since its inception. They have spent rigorous hours researching and exploring the diverse topics which have resulted in the successful publishing of this book. They have passed on their knowledge of decades through this book. To expedite this challenging task, the publisher supported the team at every step. A small team of assistant editors was also appointed to further simplify the editing procedure and attain best results for the readers.

Apart from the editorial board, the designing team has also invested a significant amount of their time in understanding the subject and creating the most relevant covers. They scrutinized every image to scout for the most suitable representation of the subject and create an appropriate cover for the book.

The publishing team has been an ardent support to the editorial, designing and production team. Their endless efforts to recruit the best for this project, has resulted in the accomplishment of this book. They are a veteran in the field of academics and their pool of knowledge is as vast as their experience in printing. Their expertise and guidance has proved useful at every step. Their uncompromising quality standards have made this book an exceptional effort. Their encouragement from time to time has been an inspiration for everyone.

The publisher and the editorial board hope that this book will prove to be a valuable piece of knowledge for students, practitioners and scholars across the globe.

Index

T

V

Campus